Blockchain in the Global South

Nir Kshetri

Blockchain in the Global South

Opportunities and Challenges for Businesses
and Societies

Nir Kshetri
University of North Carolina
Greensboro, NC, USA

ISBN 978-3-031-33943-1   ISBN 978-3-031-33944-8   (eBook)
https://doi.org/10.1007/978-3-031-33944-8

This Palgrave Macmillan imprint is published by the registered company Springer Nature
Switzerland AG
The registered company address is: Gewerbestrasse 11, 6330 Cham, Switzerland

CONTENTS

LIST OF FIGURES

LIST OF TABLES

CHAPTER 1

Blockchain in the Global South: Key Issues and Current Status

Abstract This introductory chapter offers an analysis of blockchain's potential to bring economic, political, and social transformations in the Global South. It highlights some of the key applications and uses of this technology in developing countries and takes a look at some of the visible effects. It argues that in many ways, blockchain has a much higher value proposition for the Global South than for their rich counterparts in the developed world. It also gives an overview of the uses of blockchain in promoting transparency and reducing frauds and corruption, *reducing friction* and costs of property registration, promoting efficiency in international B2B trade, and increasing access to trade and supply chain finance, reducing costs, and increasing efficiency in international payment systems. Other key uses in *finance, banking, Insurance, and* blockchain-based *digital identity* are also discussed.

Keywords Costs of networking · Costs of verification · G20 · Marine fishing industry · Permissioned blockchain · Slavery

1.1   INTRODUCTION

Blockchain is considered to have the potential to cause significant economic, political, and social transformations in economies in the Global South (GS). Blockchain affects economic, social, and political outcomes in the developing world through many direct and indirect pathways. The first of blockchain's direct benefits is the potential reduction of corruption and fraud. For instance, blockchain can empower donors. It can ensure that donations reach their intended recipients. To take an example, donors can buy electricity for a South African school using bitcoin. A blockchain-enabled smart meter makes it possible to send money directly to the meter. There are no organizations involved in distributing funds. Donors can also track the electricity being consumed by the school to calculate the power of their donations (Higgins, 2016c). This program was launched by the South African bitcoin startup Bankymoon via the crowdfunding platform Usizo, to allow public schools to use blockchain to crowdsource utility credits (Mulligan, 2015).

Increased efficiency and reduced transaction costs constitute a second benefit: there is no third party or central body involved. That is, blockchain transactions are conducted by the concerned parties themselves. There are already some signs of blockchain-led disintermediation in international remittances and international trade.

In many ways, blockchain has a much higher value proposition for economies in the GS than for the Global North. This technology has the potential to make up for the lack of effective formal institutions—rules, laws, regulations, and enforcement—in developing economies. These economies are also in desperate need of improving administrative aspects, such as maintaining standards, monitoring, and enforcing compliance. Blockchain technology is perfectly suited to address these issues.

There are different mechanisms that lead to such benefits, which can be better understood by looking at a blockchain-led reduction in the costs of verification and networking (Catalini & Gans, 2019). Regarding the cost of verification, blockchain makes it possible to verify information about past transactions and the current ownership of a digital asset. As to blockchain's effect on reducing the cost of networking, various parties can start a self-sustaining process and operate a marketplace. It is not necessary to assign control to a centralized body because blockchain can verify the state at a low cost. Economic incentives can be targeted to reward valuable activities from a network perspective, including the contribution of resources to operate and scale up the network and to secure a decentralized stage. The digital marketplaces that result from such collaborations allow participants to make joint investments to create shared digital assets (Catalini & Gans, 2019).

To illustrate these points, consider the example of blockchain-based solutions to fight global slavery. According to studies conducted by the International Labour Organization, the Walk Free Foundation, and the International Organization for Migration, in 2016, 40.3 million people were estimated to be living in modern slavery—forced to work under threat against their will or living in a forced marriage—70% of whom were women and girls (Walk Free Foundation, 2018). Many of the products that these people produce are then exported to rich countries. In 2018, the G20 countries imported $354 billion worth of products at risk of having been produced by forced labor. This is an extremely sad situation because it continues despite Western brands' efforts over three decades to address issues related to forced labor, bondage, sweatshops, and other abuses in their supply chains (Bengtsen, 2020).

The huge marine fishing industry exhibits a high propensity to use "slave" or grossly underpaid labor because of the lack of clear regulations and enforcement mechanisms. Migrant workers especially are exploited. For instance, in 2014, 82% of 172,430 fishermen employed by the Thai fishing industry were migrant workers, mainly from Cambodia and Myanmar. Most workers in seafood-processing plants are also migrants who often arrive there after falling prey to recruiters promising well-paying jobs in Thailand. However, they are paid about 25% less than Thailand's minimum wage. The migrant workers often sign a contract in their home country, but their contracts change when they arrive in the host country to begin work (Business Fights Poverty, 2020). Unlike local Thai workers, they cannot join unions and do not have other protections that Thai workers are entitled to (Nicholl, 2019).

Some initiatives are expected to improve this issue. The blockchain solutions provider Diginex has been working with the International Organization for Migration and the antislavery nongovernmental organization Mekong Club to ensure ethical recruitment of migrant workers by increasing transparency of workers' contracts. The British embassy in Bangkok partly funded the pilot phase of the project (Dao, 2019).

Blockchain-based mobile app eMin (https://www.eminproject.com) is used to store copies of employment contracts for workers in this sector. Employment contracts and related data are stored on the Ethereum blockchain. Workers can access their contracts, which allows them a basis for claiming the rights and benefits they were offered at the time of recruitment (Business Fights Poverty, 2020).

The eMin pilot started in February 2019 with Verifik8, a data intelligence and analytics provider for agribusiness suppliers, at a shrimp farm in Phuket, Thailand (Business Fights Poverty, 2020). In October 2019, Diginex signed an agreement with Verifik8 to integrate eMin into the latter's existing monitoring tools. As of October 2019, Verifik8's farming monitoring tools called Blue 8/Green 8 were used by 5,000 workers in Thailand (Dao, 2019). Diginex plans to expand into different sectors in other Southeast Asian nations, as well as in Bangladesh and Bahrain. The following sections focus on how blockchain can address some of the key challenges facing the GS. This chapter also offers an overview of enablers and barriers to implementing blockchain in economies in the GS.

1.2   BLOCKCHAIN: SOME BACKGROUND, CONCEPTS, AND FACTS

In this section, we define and make some comments about blockchain and related concepts (Table 1.1). Blockchain can be viewed as a decentralized ledger that maintains digital records of a transaction simultaneously on multiple computers. After a block of records is entered into the ledger, the information in the block is mathematically connected to other blocks. In this way, a chain of immutable records is formed (Yaga et al., 2018). Due to this mathematical relationship, the information in a block cannot be changed without changing all blocks. Any change would create a discrepancy that is likely to be noticed by others (Kshetri, 2018).

Blockchains can be permissioned (e.g., Hyperledger Fabric) or permissionless (e.g., Bitcoin and public Ethereum). Permissioned blockchains can be designed to restrict access to approved actors such as supply chain partners. In a way, permissionless blockchains are like a shared database. Everyone can read everything. However, a user cannot control who can write.

Implementing smart contracts is among blockchain's most transformative applications. Smart contracts execute automatically when certain conditions are met. A smart contract assures a party with certainty that the counterparty will fulfill the promises.

The first blockchain to implement smart contracts was Ethereum. Smart contracts are installed in each node of the Ethereum network. While Bitcoin stores data related to transactions, Ethereum stores

Table 1.1   Explanation of major terms used in the book

Term	Explanation
Blockchain	A decentralized ledger that maintains digital records of a transaction simultaneously on multiple computers
Cryptocurrency	A cryptocurrency functions like money, which means that it defines value, serves as a value transfer, and can be used for making and receiving payments. Such currencies are on the blockchain and encrypted using cryptography
Ethereum	The Ethereum network is a public blockchain-based open software platform, in which each node can be discovered by and known to other nodes in the network. It has its own cryptocurrency known as Ether
Ethereum Gas	A fraction of an Ethereum token used by a smart contract to pay for the miners' efforts to secure the transaction on the blockchain
Hyperledger Fabric	It is an open-source blockchain platform from The Linux Foundation, which is provided by IBM as "Blockchain as a Service." It is targeted at businesses. Hyperledger facilitates smart contracts by connecting all relevant parties together. Fabric is a type of private or permissioned blockchain. Some organizations or government agencies "own" the nodes, who permit the nodes to communicate with each other. Identities and roles of members are known to other members
Permissioned blockchain	In a permissioned blockchain, nodes or users are not publicly discoverable. The permission to create smart contracts may also be restricted to approved actors
Permissionless blockchain	A permissionless blockchain can allow anyone to join the network and participate in block verification to create consensus and create smart contracts. Some examples include the Bitcoin and Ethereum blockchains
Smart contracts	Smart contracts execute automatically when certain conditions are met. Computerized protocols and user interfaces are used to execute a contract's terms (Szabo, 1994) and to "formalize and secure relationships over public networks" (Szabo, 1997)

diverse types of data such as those related to finance, industry, legal, personal information, community, health, education, and governance. These data can be accessed and used by computer programs known as decentralized applications (dApps) that run on Ethereum. Software developers can choose their own "rules" for ownership, transaction formats, and other aspects (https://www.stateofthedapps.com/whats-a-dapp). Ethereum can thus be customized to offer unique solutions to special needs. It is mainly used to develop B2C applications. In Ethereum, computers connected in an open and distributed network provide the processing power needed to run a smart contract. The computers in the network also verify and record transactions in the blockchain.

The owners of the computers are awarded Ether tokens for their contributions. Ethereum can be viewed as the first shared global computer. Bitcoin, on the other hand, is considered to be the first accounting ledger that can be shared globally (MIT Technology Review, 2017). Ethereum needs what is referred to as Ethereum Gas in order to execute transactions or smart contracts.

Characteristics of Blockchain

Three key characteristics of blockchain have been identified—decentralization, immutability and cryptography-based authentication (Kshetri, 2018).

Decentralization
Blockchain's value proposition is arguably embedded in the decentralization feature. By supporting decentralized models, **blockchain** can make sustainability-related activities more transparent and hence help produce trust. Blockchain eliminates the need for a trusted third party in the transfer of value and thus enables faster, less expensive transactions. Even those who are skeptical of the potential of blockchain in many other fields and applications are optimistic about its trust-producing capabilities (Hackett, 2017).

Immutability
The term immutable comes from object-oriented programming, in which data structure and operations or functions that can be applied are defined by programmers. Immutable means that once an object has been created and is recorded in a software code, it cannot be modified (Tschantz & Ernst, 2005). Blockchain-based transactions are thus indelible and cannot be forged. The immutability feature makes transactions on blockchain auditable, which can improve transparency. A party can be given controlled access to relevant data. For instance, blockchain's distributed ledger model would allow regulators and authorities to access key data and information related to sustainability (Till et al., 2017).

Cryptography-Based Authentication
To ensure that only authorized users can access the information, blockchain systems use cryptography-based digital signatures in order to verify the identities of participants. Users sign transactions with a private key. Such a key is generated when a user creates an account. The private key is typically a very long and random alphanumeric code. Using complicated algorithms, blockchain systems also create public keys from private keys. Public keys make it possible to share information. This feature makes it possible to measure and track sustainability-related outcomes. For instance, if a coffee retailer claims that living wages are being paid to coffee farmers, the accuracy and truthfulness of such claims can be assessed by checking the payments to digital wallets that are assigned to the farmers.

How Blockchain Works

In order to illustrate how blockchain works, we consider the deployment of the technology in the diamond industry. To take an example, South Africa's De Beers Group has launched a GemFair program to log diamonds produced by artisanal and small-scale miners (ASMs) (Fig. 1.1). In the first phase of the program, De Beers trained ASMs at 16 mine sites in Sierra Leone. The training program focuses on digitally tracking diamonds throughout the supply chain. By April 2019, De Beers extended the pilot to 38 additional sites (Jamasmie, 2019).

The goal is to make sure that diamonds that originated in conflict zones do not enter the supply chain. ASMs are required to identify and manage key risks defined in a due diligence guide by the Organisation for Economic Co-operation and Development to participate in the program (Gemfair, 2019). Among the major requirements is that ASMs identify the worst forms of child labor and address them. Compliance is ensured through first-party (e.g., a member completes a self-assessment workbook provided by GemFair), second-party (GemFair's biannual site monitoring), and third-party (assessment of a sample of sites twice a year) verifications.

De Beers' blockchain solutions track diamonds as they move from the mine to cutter and polisher and then jeweler. Each organization involved in this traceability can use a smartphone or other device to sign into a blockchain platform (Fig. 1.1). De Beers' program records the GPS locations for each diamond found. The diamond is then placed in a

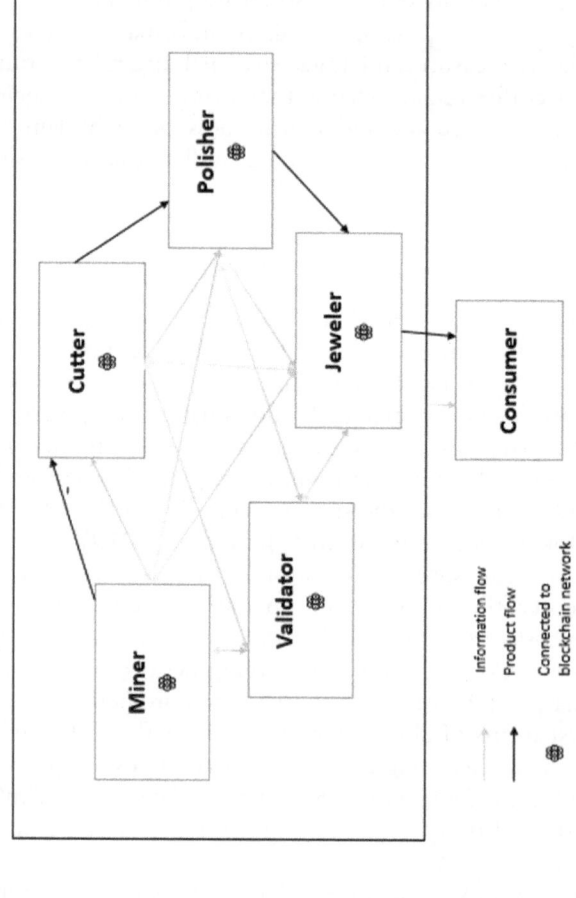

Fig. 1.1   An illustration of the use of blockchain to trace a diamond in a supply chain

tamper-proof bag with a QR code. GemFair provides a tablet for a participating mine to log into the GemFair app, which can also function offline. However, the tablet must be connected to the internet to store production records in the GemFair system. After this step, the raw diamonds move on to the supply chain's next stage. All relevant participants receive information about all transactions (Kshetri, 2022a).

1.3   Economic Prosperity and Poverty

There are many sources of underdevelopment, which include colonialism, dependence on commodities (Sindzingre, 2012), ethnic tension, corruption, exploitation, lawlessness, and political violence (Murshed, 2002). This section focuses on institutional environments in order to understand the roles of blockchain in addressing key challenges associated with such environments. Poor countries mostly lack good institutions that ensure strict enforcement of property rights, have the ability to deal with corrupt practices effectively and provide equal opportunity to all members of society (Acemoglu, 2003).

The Lack and Poor Enforcement of Property Rights

According to a 2011 report of the Food and Agriculture Organization (FAO) and Transparency International, in over 61 countries, weak governance led to corruption in land occupancy and administration. Corruption varied from small-scale bribes to the abuse of government power at the national, state, and local levels.

Enforcement of property rights increases incentives to invest and provides resources for individuals to get out of the poverty trap. Clear property rights would allow entrepreneurs to use the assets as collateral and thus increase their access to capital. A large proportion of poor people in the developing world lack property rights. For instance, about 90% of land in rural Africa is undocumented or unregistered. Likewise in India, the lack of land ownership remains among the most important barriers to entrepreneurship and economic development. One estimate is that over 20 million rural families in India did not own land and millions more lacked legal ownership of the lands where they built houses, lived, and worked (Hanstad, 2013). Indeed, lack of land ownership or tenure is arguably a more powerful predictor of poverty in India than caste or illiteracy (Hanstad, 2013).

Disregard the Rule of Law

In some developing economies, the rule of law is disregarded and not respected by corrupt politicians, government officials, and other powerful groups. These groups sometimes expropriate the incomes and investments of poor people or create an uneven playing field in other ways.

Less Opportunity for Marginalized Groups

Economically and socially disadvantaged groups have less opportunity to access finance, credit, insurance, and education. Thus, they cannot make investments and participate in productive economic activities. Consider insurance, for instance. In India, 86% of rural populations and 82% of urban populations lack health insurance (Bansal, 2016).

Regarding access to finance, in China, SMEs account for 70% of GDP but have access to 20% of financial resources. However, 89% of SMEs face difficulty satisfying banks' loan requirements. Small borrowers often lack sufficient collateral required by most traditional banks (Kshetri, 2016, 2019).

Unavailable financing is a critical barrier faced by most entrepreneurs. For instance, despite high-interest rates, demand for credit exists in most developing economies. Banks in the Democratic Republic of Congo (DRC) reject over one-third of credit and loan applications. The fact that they cannot enforce their legal rights as lenders has led to the industry's risk-averse behavior, and this is a manifestation of a broader structural problem in economies in the GS where a large proportion of the population lacks access to formal banking institutions (Kshetri, 2019).

Barriers

Poor-quality institutions lead to transaction cost barriers. Making this statement meaningful requires a more detailed discussion of transaction costs. In the context of business transactions involving two or more parties, for Douglas North, "transaction costs are... two things: (1) the costs of measuring the dimensions of whatever it is that is being produced or exchanged and (2) the costs of enforcement." (North, 1999). He goes on to say that "a lot of what we need to do is to try to measure the dimensions of what we are talking about in such a way that we can define them precisely" (North, 1999).

Many developing economies are faced with challenges in enforcing commercial contracts, social and economic rights, laws and regulations (e.g., agro-environmental), and standards (e.g., pollution). Put differently, these economies lack effective contract enforcement mechanisms (Fig. 1.2). The main goal of contract laws and their enforcement is to increase the value of contracting in order to facilitate the organization of economic activities. When there is a legal remedy for breach of contract, a party to a contract is less likely to engage in opportunistic behavior. In many GS economies, laws in general and contract laws in particular are not adequate to meet the needs of the population (Posner, 2005; Williamson, 1985). Enforcing a contract takes longer and is more costly in economies in the GS than in developed countries.

Detailed contracts can protect against a contracting partner's opportunism through the threat of legal enforcement. Most parties, however, rarely use detailed contracts in practice because of the high costs associated with drafting and enforcing them (Macaulay, 1963). This is especially an issue of concern in developing countries.

Emphasizing the importance of measurement to enforcement, North argues: "Without being able to measure accurately whatever it is you are

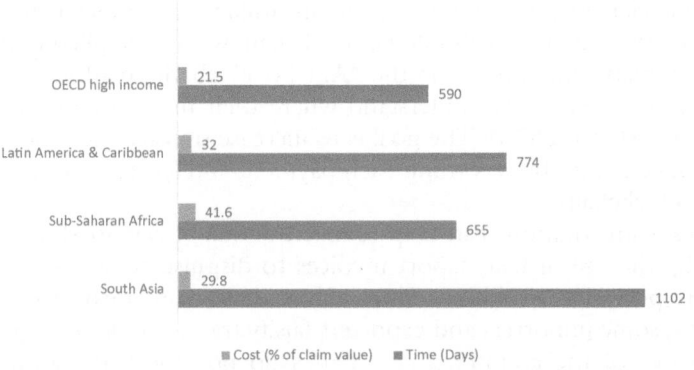

Fig. 1.2 Time and cost required to enforce contracts in different geographic regions and groups of countries (*Data source* World Bank. "Doing business measuring business regulations" 2020, https://www.doingbusiness.org/en/data/exploretopics/enforcing-contracts#)

trying to enforce, there cannot be effective enforcement, even as a possibility" (North, 1999). The technology available is among the important factors that affect the costs of measurement and enforcement and so transaction costs (North, 1999). In this, blockchain can make up for the lack of relevant institutions or the problems associated with high transaction costs.

Enforcements can be implemented at three levels: first, second, and third parties (North, 1999). Third-party enforcement mechanisms, which are often formal coercive enforcement measures by the state, have been relatively ineffective in developing economies[42]. Blockchain has the potential to strengthen the government's enforcement powers and sanctions against individuals or organizations that breach regulations.

1.4   NOTABLE BLOCKCHAIN APPLICATIONS IN THE GS

In this section, we discuss some key current applications and future prospects of blockchain.

Promoting Transparency and Reducing Risk

Blockchain can help achieve transparency in various settings. In 2016, Ant Financial, Alibaba's online payments affiliate, announced the launch of blockchain payment technology. Blockchain was first applied to Alipay's donation platform. Donors on the "Ant Love" charity platform can track transaction histories and understand where their funds go and how they are used (Kshetri, 2023). The goal is to increase transparency and provide a trust mechanism by recording each payment and spending of donations on the blockchain.

Blockchain solutions can help reduce fraudulent activities. To take an example, the use of fake export invoices to disguise cross-border capital flows is pervasive in China. Since China has maintained strict capital controls, some importers and exporters falsify transactions to move capital in and out of the country. Many banks do not check the authenticity of trade documents (Shengxia, 2014). From April to September 2014, China found $10 billion in fake transactions (Shengxia, 2014). Some major fraud cases were in Qingdao, the world's seventh-busiest port, where firms had used fake receipts to secure multiple loans against a single cargo of metal (Smith, 2015).

The Qingdao fraud involved 300,000 tons of aluminum, 20,000 tons of copper, and 80,000 tons of aluminum ingots (Shengxia, 2014). As a result of this scandal, Chinese banks charge higher interest rates and are less likely to provide collateral financing (Smith, 2015). Blockchain, though, can arguably stop such scandals.

Recent high-profile fraud has increased blockchain's attractiveness. The British multinational banking and financial services company Standard Chartered lost about $200 million from Qingdao fraud. Standard Chartered teamed up with DBS Group and Singapore's Infocomm Development Authority to develop a blockchain-based platform. Standard Chartered is a participant in blockchain-based trade finance platforms such as eTradeConnect and Bay Area Trade Finance Blockchain Platform.

Some financial technology companies aim to address this problem by providing blockchain-based systems for the reimbursement process (known as *fapiao* in Chinese). In August 2018, Tencent piloted a blockchain-based feature using WeChat Pay data to inform employers of employees' purchases. Employees can use the system to automatically send transaction data to employers for reimbursement. The feature is expected to bring efficiency to the corporate expense reimbursement process and reduce fraud and tax evasion (Idowu, 2019).

Currently, reimbursement requires that merchants issue different receipts for purchases with the employer's taxpayer ID and other information. Merchants manually enter this information to generate receipts and process additional paperwork (Wang, 2018). In December 2018, Tencent announced that qualified merchants can use WeChat's blockchain-backed e-invoices (Wang, 2018).

Reducing Frictions in Property Registration

Blockchain can play a major role in improving the protection of property rights (Kshetri, 2019). Some potentially high-impact uses of blockchain are likely to be in improving property registry (Swan, 2015), and also securely managing land records and land tenure (Zwitter & Despiaux, 2018). A blockchain-based property registry system can reduce title fraud and guarantee title protection (Themistocleous, 2018). Blockchain can reduce friction, conflict, and costs in property registration. Land-titling systems for blockchain applications can be carried out in an inexpensive way (Kshetri, 2022b).

Various benefits of blockchain in developing a national system for property management have been suggested (Gabison, 2016). They include the elimination of paperwork, reduction of fraud, and increase in speed with which transactions can be conducted (Lantmateriet, 2016).

Promoting Efficiency and Access to International Trade and Financing

The global trade finance market, valued at $18 trillion, is likely to be transformed by blockchain's disintermediation and other efficiency measures. A current challenge is that there is a big gap between the demand and supply of trade financing. According to the Asian Development Bank (ADB), the global trade finance gap was $1.5 trillion, or 10% of merchandise trade volume in 2018. This gap is expected to increase to US$2.4 trillion by 2025[78].

First, the global trade finance market relies on paper documentation for most processes. A typical cross-border transaction involves many parties. A letter of credit (LC)—a promise to pay for goods if certain conditions are fulfilled—is sent to the exporter by the importer's bank. After receiving the LC, the exporter ships the goods. The bank faces the risk that the importer may be unable or unwilling to pay. The exporter then presents proof of shipping to get financing from its bank, which is paid directly by the importer's bank. Estimates suggest that, on average, a single cross-border trade transaction involves the exchange of 36 documents (Fletcher, 2019) to 40 documents (Lamoureux & Evans, 2011). As many as 240 copies of documents need to be exchanged among various parties, such as financiers, logistics providers, customs officers, and warehouse managers (Fletcher, 2019).

Paper-based methods such as letters of credit and factoring account for about US$5 trillion of annual trade worldwide (Allison, 2016). It costs 1–3% of a trade's value to buy an LC. Paper documents need to be physically exchanged, an extremely slow process that became especially apparent during the COVID-19 pandemic. Documents such as letters of credit, bills of lading, and invoices are normally carried in the cargo holds of passenger aircraft. Most passenger flights, though, could not operate during the COVID-19 pandemic. Millions of documents related to cross-border trade transactions were forced to move to alternative means, such as ships, to reach their destinations. However, many could not be delivered to banks, which were closed. Due to the emergency nature of the situation, many banks started accepting scanned signatures and documents. While electronic documents show superior performance in terms of speed, their proneness to fraud has been a big concern among banks and other players. Blockchain has clear security benefits in this regard (Tan, 2020).

Factors are key intermediary players in the global trade finance market. They offer money to exporters. Based on promised future payments, exporters borrow from factors. Exporting firms make an outright sale of accounts receivable to factors in order to maintain liquidity. For instance, a Chinese exporter selling to Walmart can take the invoice for those goods to a factor, which pays the exporter right away. For a $100 invoice, the factor may pay as little as US$90. The upshot is that buyers such as Walmart pay higher for goods they buy from sellers in the developing world. The global factoring market is estimated at over US$2 trillion annually (Tan, 2020).

Several companies have created blockchain-based products to address inefficiencies in B2B trade and supply chain financing (PRNewswire, 2016). The products are expected to eliminate intermediaries and financiers. For example, buyers and sellers agree on the terms of a deal, then blockchain tracks and manages the transaction from start to finish. In March 2017, China's internet financial services company Dianrong and FnConn, the Chinese subsidiary of the Taiwanese electronics manufacturer Foxconn, launched Chained Finance, China's first blockchain platform for supply chain finance (CCN.com, 2020). Electronics, auto manufacturing, and clothing companies facing difficulties in accessing supply chain financing are the test markets for Chained Finance (Kshetri, 2017a). Instead of charging suppliers, Chained Finance charges peer-to-peer (P2P) lenders a fee to access the system. Using the platform, nonbank lenders can make direct loans in supply chains worldwide (Peterson, 2019). Before launch, the two companies had successfully completed a pilot project and proof of concept to secure US$6.5 million in funding for Chinese SMEs. Different levels of suppliers are expected to be connected to the system of Chained Finance, and the company aims to expand to other developing economies. As of early 2020, more than 20 electronics suppliers were being paid on Ethereum-based cryptocurrency. The company reported that financing costs reduced from 24% per year to 10% and the time needed to get funding from seven days to one (Forbes, 2020).

Finance, Banking, and Insurance

In the insurance sector, blockchain may provide risk managers with an effective way to protect individuals and companies from uncertain losses or catastrophes. Insurance and derivatives can be used to control or minimize risk factors associated with unpredictable events. By supporting decentralized insurance models, blockchain may make derivatives more transparent. A meaningful risk management process can be designed using reputational systems based on individuals' social and economic capital and online behavior. Blockchain-based insurance, for example, is connected to big data, the Internet of Things (IoT), and health trackers to ensure better pricing and risk assessment (Ramada-Sarasola, 2016).

The IoT makes it easier for cars, electronic devices, and home appliances to have their own insurance policies. Using blockchain, these devices can be registered and their insurance policies administered by smart contracts. Damage is automatically detected, which triggers the repair process, claims, and payments. Payouts are made against the insurable event, and the policyholder does not have to make a claim. The insurer does not need to administer claims. The costs of claims processing drop close to zero. Even more important, there is less likelihood of fraud (Kshetri, 2017b).

Some blockchain-based insurance products have been launched in the GS. Take an example, the Mexican mobile payment platform Saldo.mx launched a microinsurance service, Consuelo, which allows users to buy blockchain-powered health and life insurance policies. The target groups are Mexican living in the country as well as in the diaspora (Valenzuela, 2015).

Some innovative blockchain solutions have helped drive financial inclusion. Blockchain-based solutions make peer-to-peer lending possible by directly connecting lenders and borrowers, thereby eliminating the need for intermediaries. Consider Kiva. The company does not make direct loans. While some investors mistakenly think that Kiva offers direct person-to-person connections, it actually works with local microfinance institutions (MFIs). Kiva conducts audits of its field partners to ensure that low-income groups are not exploited. However, due to high overhead costs and other inefficiencies, Kiva field partners charge exorbitantly high-interest rates. For instance, according to Female Founder Stories, which publishes interviews and insights from the female alumni of the accelerator YCombinator, a Kiva field partner in Senegal was reported to charge an interest rate of 40%. Such loans could be made more affordable by eliminating intermediaries such as Kiva field partners.

Central bank digital currencies (CBDCs), many of which are based on blockchain, can also promote financial inclusion (Allen et al., 2020). The International Monetary Fund argues that CBDCs offer great promise for reaching marginalized groups (Chatenay, 2021). A retail CBDC system—in which a central bank issues digital currency directly, without the need for traditional bank accounts—could be a game changer in eliminating poverty. This can be achieved through the establishment of an inclusive digital payment ecosystem and the creation of financial data identities. For instance, individuals can have a CBDC account on the central bank's ledger. A digital wallet application linked to the account through application program interfaces (APIs) can allow users to access their accounts and engage in transactions (Raghuveera, 2020). In China, for instance, since the digital yuan is highly traceable, the country's central bank, People's Bank of China (PBOC), can monitor the flow of money in the economy (Huang, 2021). This allows the government to deliver targeted programs to improve the well-being of high-risk groups such as SMEs and low-income households.

Blockchain-Based Digital Identity

According to the World Bank's ID4D database, 1 billion people lack any form of identification. An additional 3.4 billion people have some type of identification but lack the ability to use it in the digital world. Identity management is thus a big issue. In financial institutions, the ability to prove someone is who they say they are is very important for increasing the accuracy of risk assessment and reducing fraud. Potential borrowers in many Global South economies cannot prove who they are, which is among the main reasons many low-income people lack access to financial services (Kshetri, 2020, 2023).

Blockchain can play a major role in creating secure digital identities. Since ID cards in many countries are paper, which can be easily forged, blockchain solutions have significant potential to reduce fraudulent activities. Blockchain increases the ability to get secure and authentic identity proof at a low cost. BanQu's blockchain-based verifiable digital identity for the GS is expected to help marginalized groups establish ownership, business assets, and production values and help them engage in economic transactions.

Some economies in the GS have started working on blockchain-based ID. In August 2019, Sierra Leone launched the blockchain-based National Digital Identity Platform developed by Kiva. Kiva worked with the UN Capital Development Fund and the UN Development Program to develop the platform. Kiva's blockchain protocol aims to address two major barriers that hinder low-income people's access to financial services: formal identification and verifiable credit history (Kshetri, 2020).

1.5   Blockchain Diffusion

Activities of Foreign Multinationals

Blockchain companies from the industrialized world are making inroads into developing economies. In 2017, the British-Dutch multinational consumer goods company Unilever announced that it had teamed up with British supermarket chain Sainsbury, packaging company Sappi and three global financial services companies—BNP Paribas, Barclays, and Standard Chartered—to develop a blockchain platform to track sustainability practices in its supply chains. The initial year-long project started with $700,000 in funding from the U.K.'s Department for International Development and private sources to track two categories of products: tea supplies used by Unilever and Sainsbury's and wood fibers in certain Sappi packaging solutions. The plan was to start with a system to track and verify contracts for farmers in Malawi who supplied tea to Unilever and Sainsbury. Financial incentives were offered to the tea farmers for feeding social or ecological data into the blockchain system (*Sappi*, 2019).

The initiative is expected to reach up to 10,000 farmers. The group announced that preferential pricing would be applied to farms that engage in sustainable farming methods, which increase harvests without using more land. The banks would finance farms that utilize sustainable farming methods (Clancy, 2017).

Provenance, the blockchain-enabled fintech startup Halotrade, and the real estate development company Meridia developed the "Trado model." It facilitates data-sharing among producers, consumers, and supply chain players (*Provenance* nd). The Trado model was first piloted in the tea supply chain in collaboration with the Lujeri Tea Estate in Malawi. It made farming data from smallholder farmers directly accessible to Unilever. Data related to tea leaf production, social impact, and sustainability credentials were recorded in the Provenance platform. Using data related to the availability of goods, Unilever supported the release of payment earlier. The 18-month pilot found that Unilever and Sainsbury's could increase supply chain transparency with the data (George, 2019). It reduced the period of expensive financing for upstream partners. The transactions take place using a bank's regular supply chain finance process, which results in minimal disturbance to the banks' businesses (finextra.com, 2019).

The resulting savings were invested to fund projects such as a field school for local farmers to provide training in sustainable practices. Local NGOs confirmed payment distributions to the school and recorded impact progress on the blockchain (*Provenance*, n.d.).

Local Entrepreneurial Activities

In some economies in the GS, entrepreneurial activities in blockchain are rapidly rising. For instance, as of 2021, Africa had 60 active cryptocurrency exchange platforms that focused on diverse activities such as P2P transactions and trade financing (The Baobab Network, 2021). To take an example, the Kenyan fintech firm Pezesha, which specializes in micro-, small-, and medium-sized enterprise (MSME) credit scoring and loan origination, developed a crypto-based solution to allow global lenders to invest in Africa. Foreign lenders can send stablecoin in U.S. dollars for conversion to Kenyan shillings. As of March 2022, Pezesha had facilitated 3,751 loans in Kenya and 344 in Ghana (Ndemo, 2022). Since MSMEs bring benefits to the poor (Asikhia, 2010), startups such as Pezesha play a key role in poverty alleviation.

Favorable Policies for Developing Blockchain Ecosystems

Some economies in the GS have formulated and implemented favorable policies for developing blockchain ecosystems. For instance, China's 13th Five-Year Plan on Informatisation from 2016 to 2020 has focused on advanced technologies, such as big data, AI, and blockchain. The plan has listed blockchain technologies among the main development directions (Coleman, 2016). In 2016, the Chinese government also published a white paper that emphasized the importance of "quick response and reasonable planning" to influence the international standardization of blockchain. The report set a deadline of April 2017 for pilots using blockchain standards developed by the group (Wong & Huang, 2016).

Likewise, the No. 1 Central Document, released in 2017, has given due recognition to agricultural innovation as a priority area, especially quality supervision and standards. Note that the "No. 1 Central Document" is the first policy statement released by the central government in a given year. This document is seen as an indicator of the government's policy priorities.

Likewise, the government of Mauritius has devoted resources to developing the blockchain ecosystem. Mauritius has collaborated with the private sector in the country and internationally. The government and the blockchain software technology firm ConsenSys were exploring potential collaboration to establish the foundational elements of a blockchain ecosystem, including "know your customer," or KYC, rules, digital identity, and title registries. In the subsequent phase, it plans to help the country to build a talent pool of developers, entrepreneurs, executives, and regulators to further enrich the ecosystem (Stanley, 2017). The Mauritius government and ConsenSys were reported to be exploring the possibility of opening a ConsenSys Academy in Mauritius (NewsBTC, 2017). ConsenSys Academy Dubai's first class of Ethereum blockchain developers graduated in 2017.

1.6   Chapter Summary and Conclusion

Among the most attractive features of blockchain is that once a record is created, it is almost impossible to be tampered with or forged. Blockchain will thus make data secure. Transactions can also be conducted to achieve any degree of privacy or openness. Some application areas include land registration and donation and payment tracking. Cryptocurrencies as an interoperable system can more easily convert various currencies and facilitate cross-border trade.

Programs such as Kiva's blockchain-based IDs are a first step to improving access to finance for low-income populations. True decentralization will be complete when impact investors and philanthropic funders can directly reach low-income groups with cryptocurrencies.

Blockchain helps prevent corrupt officials from engaging in fraudulent activities. Organizations can make sure that their business partners play by the rules. Services providers can make sure that people are who they say they are when they enroll and participate in various services. These features are of special interest in developing economies, which can lack effective and trustworthy institutions.

Blockchain is especially likely to make contract enforcement more efficient and effective. For instance, a blockchain-based life or health insurance contract can be a powerful tool: it is possible to automatically activate a policy based on the diagnosis. For instance, if a diagnosis indicates the existence of a triggering condition for the policy that is written in the smart contract, the information is fed to the blockchain. The smart contract automatically authorizes payments based on the policy. Smart contracts can also act as a warranty for a down payment to the medical service provider, and there is no need to have a previous contractual relationship between the medical service provider and the insurance company. In this way, smart contracts drastically reduce administrative costs.

References

Acemoglu, D. (2003). Root Causes: A historical approach to assessing the role of institutions in economic development. *Finance & Development*, 27–30. http://isites.harvard.edu/fs/docs/icb.topic637539.files/Acemoglu.pdf

Acheson, N. (2017, June 26). *Counting chickens: Can blockchain restore trust in China's food supply?* Coindesk. http://www.coindesk.com/counting-chickens-can-blockchain-restore-trust-in-chinas-food-supply/

Allen, S., et al. (2020). *Design choices for central bank digital currency: policy and technical considerations.* Brookings. https://www.brookings.edu/wp-content/uploads/2020/07/Design-Choices-for-CBDC_Final-for-web.pdf

Allison, I. (2016). Skuchain: Here's how blockchain will save global trade a trillion dollars. *International Business Times*. Available at https://www.ibtimes.co.uk/skuchain-heres-how-blockchain-will-save-global-trade-trillion-dollars-1540618

Asikhia, O. U. (2010). SMEs and poverty alleviation in Nigeria: Marketing resources and capabilities implications. *New England Journal of Entrepreneurship, 13*(2), 57–70. https://doi.org/10.1108/NEJE-13-02-2010-B005

Bansal, S. (2016). Health cover: too little, too scarce. *The Hindu*. http://www.thehindu.com/sci-tech/health/policy-and-issues/health-insurance-in-india-too-little-too-scarce-reveal-national-sample-survey-data/article8462747.ece

Barboza, D. (2013, August 3). Coin of realm in China graft: Phony receipts. *New York Times*. https://www.nytimes.com/2013/08/04/business/global/coin-of-realm-in-china-graft-phony-receipts.html

Bengtsen, P. (2020, August 28). "Why are monitory democracies not monitoring supply chain slavery? *Global Policy Journal*. https://www.globalpolicyjournal.com/blog/28/08/2020/why-are-monitory-democracies-not-monitoring-supply-chain-slavery

BoostVC. (2016). *5 ways that insurance will be disrupted by the blockchain*. https://medium.com/boost-vc/5-ways-that-insurance-will-be-disrupted-by-the-blockchain-8ffc33674713#.4ng8nof0w

Business Fights Poverty. (2020, January 21). *Detecting modern slavery in the supply chain*. Business Fights Poverty. https://wordonthestreets.net/Articles/560874/Detecting_modern_slavery.aspx

Catalini, C., & Gans, J. S. (2019). *Some Simple Economics of the Blockchain* (Rotman School of Management Working Paper 2874598).

CCN.com. (2020). *Chinese fintech firms launch Blockchain Supply Chain Finance Platform*. Available at https://www.cryptocoinsnews.com/chinese-fintech-firms-launch-blockchain-supply-chain-finance-platform/

Chatenay, V. (2021). Facebook-backed Diem has cleared regulatory hurdles to finally launch in Q1. *Business Insider*. https://www.businessinsider.com/facebook-digital-currency-to-finally-launch-q1-2021-2

Cheney, C. (2019). *In Sierra Leone, new Kiva Protocol uses blockchain to benefit unbanked*. Devex. www.devex.com/news/in-sierra-leone-new-kiva-protocol-uses-blockchain-to-benefit-unbanked-95490

China Daily. (2020, September 18). World's top 10 patent holders by blockchain inventions. *China Daily*. https://www.chinadaily.com.cn/a/202009/18/WS5f63e5c0a31024ad0ba7a3fa_1.html

Clancy, H. (2017, December 13). *Unilever teams with big banks on blockchain for supply chain*. Greenbiz. https://www.greenbiz.com/article/unilever-teams-big-banks-blockchain-supply-chain

Coleman, L. (2016, December 29). *China to support blockchain development under new five-year plan*. Cryptocoinsnews. https://www.cryptocoinsnews.com/china-support-blockchain-development-new-five-year-plan/

Crowdfund Insider. (2020). https://www.crowdfundinsider.com/2020/03/158535-peoples-bank-of-china-acquires-4-7-million-in-funding-to-further-develop-blockchain-based-trade-finance-platform/

Dao, T. (2019). *Companies ink deal to use blockchain for protecting Thai aquaculture sector workers*. Seafood Source. https://www.seafoodsource.com/news/aquaculture/companies-ink-deal-to-use-blockchain-for-protecting-thai-aquaculture-sector-workers

Dovi, E. (2011). *Boosting domestic savings in Africa*. http://www.un.org/africarenewal/magazine/october-2008/boosting-domestic-savings-africa

finextra.com. (2019, September 17). *Banks back pilot bidding to unlock finance for sustainability in supply chains*. finextra.com. https://www.finextra.com/newsarticle/34401/banks-back-pilot-bidding-to-unlock-finance-for-sustainability-in-supply-chains

Fletcher, L. (2019). Forget the paper trail—Blockchain set to shake up trade finance. *Financial Times*. https://www.ft.com/content/04a4fcde-dfb5-11e9-b8e0-026e07cbe5b4

Forbes. (2020). *Bitcoin's guardian angel; the 50 biggest companies in Blockchain, Forbes*. Forbes Magazine. Available at https://www.forbes.com/sites/crypto confidential/2020/02/23/bitcoins-guardian-angel-the-50-biggest-compan ies-in-blockchain/#25a0ed083fcf

Gabison, G. (2016). Policy considerations for the blockchain technology public and private applications. *SMU Science and Technology Law Review, 19*, 327–350. https://heinonline.org/HOL/LandingPage?handle=hein.journals/com lrtj19&div=19&id=&page=

Gemfair. (2019). *Artisanal and small-scale mining standard*. Gemfair. https://gemfair.com/static/files/GemFair_ASM_Requirements_2019_v2.pdf

George, S. (2019, September 17). *Blockchain-enabled supply chain sustainability scheme hailed 'successful' by business giants*. edie. https://www.edie.net/news/8/Blockchain-enabled-supply-chain-sustainability-scheme-hailed--successful--by-business-giants/

Hackett, R. (2017). *Reasons why China banned ICOs*. http://fortune.com/2017/09/05/china-bitcoin-blockchain-ico-ban/

Hager, L. M. (1972). The role of lawyers in developing countries. *American Bar Association Journal, 58*, 33–38.

Hanstad, T. (2013). *The case for land reform in India*. Foreign Affairs. https://www.foreignaffairs.com/articles/india/2013-02-19/untitled?cid=soc-twitter-in-snapshots-untitled-022013

Higgins, S. (2016a). *Republic of Georgia to develop blockchain land registry*. Coindesk. http://www.coindesk.com/bitfury-working-with-georgian-govern ment-on-blockchain-land-registry/

Higgins, S. (2016b). *How bitcoin brought electricity to a South African school*. CoinDesk. http://www.coindesk.com/south-african-primary-school-blockchain/

Higgins, S. (2016c). *Survey: Blockchain capital markets spending to reach $1 billion in 2016*. http://www.coindesk.com/capital-markets-1-billion-2016c-blockchain/

Huang, X. (2021). *China's DCEP project launches biggest digital yuan test yet*. Forkast. https://forkast.news/china-dcep-digital-yuan-pros-cons/

Huckstep, R. (2016). *What does the future hold for blockchain and insurance?* https://dailyfintech.com/2016/01/14/what-does-the-future-hold-for-blo ckchain-and-insurance

Huillet, M. (2020). *Egyptian national bank turns to blockchain to boost remittance business*. Cointelegraph. https://cointelegraph.com/news/egyptian-national-bank-turns-to-blockchain-to-boost-remittance-business

Idowu, J. (2019, July 22). *Tencent to use blockchain on WeChat for faster refunds of company expenses*. BTCNN. https://www.btcnn.com/tencent-to-use-blo ckchain-on-wechat-for-faster-refunds-of-company-expenses/

Inveen, C. (2019). *San Francisco crowdfunder Kiva sets up Sierra Leone credit database*. Reuters. www.reuters.com/article/us-leone-kiva/san-francisco-crowdfunder-kiva-sets-up-sierra-leone-credit-database-idUSKCN1VB262

Jamasmie, C. (2019). *De Beers expands pilot scheme in Sierra Leone to sell ethically sourced diamonds*. Mining.com. https://www.mining.com/de-beers-expands-pilot-scheme-sierra-leone-sell-ethically-sourced-diamonds/

Jones, A. (2016). *How blockchain is impacting industry*. International Banker. http://internationalbanker.com/finance/blockchain-impacting-industry/

Klein, B. P., & Cukier, K. N. (2009). *Tamed tigers, distressed dragon*. https://www.foreignaffairs.com/articles/asia/2009-07-01/tamed-tigers-distressed-dragon

Kshetri, N. (2016). Big data's role in expanding access to financial services in China. *International Journal of Information Management, 36*(3), 297–308.

Kshetri, N. (2017a). Will blockchain emerge as a tool to break the poverty chain in the Global South? *Third World Quarterly, 38*(8), 1710–1732.

Kshetri, N. (2017b). Blockchain's roles in strengthening cybersecurity and protecting privacy. *Telecommunications Policy, 41*(10), 1027–1038.

Kshetri, N. (2018, May 29). Blockchain could be the answer to cybersecurity. maybe. *Wall Street Journal*. https://www.wsj.com/articles/blockchain-could-be-the-answer-to-cybersecurity-maybe-1527645960

Kshetri, N. (2019). *Global entrepreneurship: Environment and strategy* (2nd ed.). Routledge.

Kshetri, N. (2020). Blockchain-based financial technologies and cryptocurrencies for low-income people: Technical potential versus practical reality. *IEEE Computer, 53*(1), 18–29.

Kshetri, N. (2022a). Blockchain systems and ethical sourcing in the mineral and metal industry: A multiple case study. *The International Journal of Logistics Management, 33*(1), 1–27. https://doi.org/10.1108/IJLM-02-2021-0108

Kshetri, N. (2022b). Blockchain as a tool to facilitate property rights protection in the Global South: Lessons from India's Andhra Pradesh State. *Third World Quarterly, 43*(2) 371–392. https://doi.org/10.1080/01436597.2021.2013116

Kshetri, N. (2023). *Fourth revolution and the bottom four billion: Making technologies work for the poor*. University of Michigan Press.

Lamoureux, J., & Evans, T. (2011). *Supply chain finance: a new means to support the competitiveness and resilience of global value chains*. Social Science Research Network (SSRN). https://papers.ssrn.com/sol3/papers.cfm?abstractid=2179944

Lantmateriet. (2016). *The land registry in the blockchain* (Blockchain Landregistry Report). http://icait.org/pdf/Blockchain_Landregistry_Report.pdf

ledgerinsights.com. (2019). *Kiva sets up Sierra Leone Blockchain ID System, Ledger Insights—Blockchain for enterprise.* Available at https://www.ledgerins ights.com/kiva-sierra-leone-blockchain-id-system/

Lemieux, V. (2016). Trusting records: Is blockchain technology the answer? *Records Management Journal, 26*(2), 110–139. https://doi.org/10.1108/RMJ-12-2015-0042

Macaulay, S. (1963). Non-contractual relations in business: A preliminary study. *American Sociological Review, 28*(1), 55–67.

Manski, S. (2017). Building the blockchain world: Technological commonwealth or just more of the same? *Strategic Change, 26*(5), 511–522. https://doi.org/10.1002/jsc.2151

MIT Technology Review. (2017). *Understand why Ethereum exists, and you'll get why it's a big deal.* https://medium.com/mit-technology-review/unders tand-why-ethereum-exists-and-youll-get-why-it-s-a-big-deal-df6765a5805d

Mulligan, G. (2015). *5 African crowdfunding startups to match.* Disrupt Arica. http://disrupt-africa.com/2015/11/5-african-crowdfunding-startups-to-watch/

Murshed, S. M. (2002). Conflict, civil war and underdevelopment: An introduction. *Journal of Peace Research, 39*(4), 387–393.

Ndemo. (2022, March 16). *Bitange The role of cryptocurrencies in sub-Saharan Africa.* Brookings. https://www.brookings.edu/blog/africa-in-focus/2022/03/16/the-role-of-cryptocurrencies-in-sub-saharan-africa/

NewsBTC. (2017). *Mauritius plans to create an Ethereum Island.* https://www.newsbtc.com/2017/07/16/mauritius-plans-create-ethereum-island/

Nguyen, A. (2019). *Thai bank targets $1 billion spinoff among its fintech units.* Bloomberg. www.bloomberg.com/news/articles/2019-11-21/thai-bank-tar gets-1-billion-spinoff-among-its-fintech-units

Nicholl, K., Wilhelm, M., & Bhakoo, V. (2019). *Almost every brand of tuna on supermarket shelves shows why modern slavery laws are needed.* The Conversation; 2019, https://theconversation.com/almost-every-brand-of-tuna-on-sup ermarket-shelves-shows-why-modern-slavery-laws-are-needed-108421

North, D. C. (1999). Dealing with a nonergodic world: Institutional economics, property rights, and the global environment. *Duke Environment, Law, and Policy Forum, 10*(1), 1–12.

Ogundej, O. (2016, May 24). *Land registry based on blockchain for Africa.* IT Web. https://itweb.africa/content/raYAyModrd4qJ38N

Partz, H. (2020, September 15). *Russia's largest bank joins blockchain trade finance platform.* Cointelegraph. https://cointelegraph.com/news/russia-s-largest-bank-joins-blockchain-trade-finance-platform

Peng, T. (2020, August 9). *More than 10,000 new blockchain companies established in China in 2020.* Cointelegraph. https://cointelegraph.com/news/more-than-10-000-new-blockchain-companies-established-in-china-in-2020

Peterson, K. (2019). *Foxconn founder: Libra can 'converge' with China's digital currency in Taiwan*. Moon Catcher. https://dailynews.bitcoindiamond.org/foxconn-founder-libra-can-converge-with-chinas-digital-currency-in-taiwan/

Posner, R. A. (2005). The law and economics of contract interpretation. *Texas Law Review, 83*, 1581–1614.

PRNewswire. (2016). *Skuchain developing blockchain solutions for $18 trillion trade finance market with funding from Amino*. DCG, and FBS Capital. http://www.prnewswire.com/news-releases/skuchain-developing-blockchain-solutions-for-18-trillion-trade-finance-market-with-funding-fromamino-dcg-and-fbs-capital-300214205.html

Provenance. *Leveraging new technologies to fund fair, sustainable smallholder farming*. Provenance. https://www.provenance.org/case-studies/unilever

PYMNTS. (2018, June 25). *AliPay, GCash launch blockchain cross-border remittance service*. https://www.pymnts.com/news/cross-border-commerce/2018/alipay-gcash-blockchain-cross-border-remittance-philippines/

Raghuveera, N. (2020, June 10). *Central bank digital currency can contribute to financial inclusion but cannot solve its root causes*. Atlantic Council. https://www.atlanticcouncil.org/blogs/geotech-cues/central-bank-digital-currency-can-contribute-to-financial-inclusion-but-cannot-solve-its-root-causes/

Ramada-Sarasola, M. (2016). *Want to get an insurer's attention? Just say blockchain*. https://www.willistowerswatson.com/en/insights/2016/06/want-to-get-an-insurers-attention-just-sayblockchain

Ripple. (2020). *Azimo and SCB runs on ripple for instant payments into Thailand*. https://ripple.com/insights/azimo-and-scb-runs-on-ripple-for-instant-payments-into-thailand/

Sappi. (2019). *Sappi teams up with major global brands to explore the potential of innovative blockchain technology in enhancing the sustainability of global supply chains*. Sappi. https://www.sappi.com/sappi-teams-up-with-major-global-brands-to-explore-the-potential-of-innovative-blockchain-technology

Shengxia, S. (2014, September 26). China uncovers $10b worth of falsified trade. *People's Daily Online*. http://en.people.cn/n3/2016/1129/c90000-9148331.html

Sindzingre, A. (2012). The impact of the 2008–2009 crisis on commodity-dependent low-income African countries: Confirming the relevance of the concept of poverty trap? *Journal of International Development, 24*(8), 989–1007.

Smith, P. (2015). *7 ways blockchain technology could disrupt the post-trade ecosystem* (Kynetix White Paper). http://www.the-blockchain.com/docs/Seven%20ways%20the%20Blockchain%20can%20change%20the%20trade%20system.pdf

Stanley, A. (2017, May 7). *Mauritius: The tropical paradise looking to become a blockchain hub*. CoinDesk. https://www.coindesk.com/mauritius-the-tropical-paradise-looking-to-become-a-blockchain-hub

Swan, M. (2015). *Blockchain: Blueprint for a new economy.* O'Reilly Media.

Szabo, N. (1994). *Smart contracts.* Unpublished manuscript.

Szabo, N. (1997). Formalizing and securing relationships on public networks. *First Monday, 2*(9).

Tan, P. (2000). *Coronavirus hastens trade finance's blockchain moment.* BBN Times. https://www.bbntimes.com/technology/coronavirus-hastens-trade-finance-s-blockchain-moment

Tapscott, D. (2016). How will blockchain change banking? How won't it? *Huffington Post.* http://www.huffingtonpost.com/don-tapscott/how-will-blockchain-chang_b_9998348.html

The Baobab Network 60 cryptocurrency-enabled fintech's—Africa market map February 7, 2021 https://insights.thebaobabnetwork.com/2021-02-cryptocurrency-africa-market-map/

Themistocleous, M. (2018). Blockchain technology and land registry. *Cyprus Review, 30*(2), 195– 202. http://cyprusreview.org/index.php/cr/article/view/579

Till, B. M., Afshar, S., Peters, A. W., & Meara, J. G. (2017). Blockchain and global health: How the technology could cut waste and reduce fraud. *Foreign Affairs.* https://www.foreignaffairs.com/articles/world/2017-11-03/blockchain-and-global-health

Tschantz, M. S., & Ernst, M. D. (2005). Javari: Adding reference immutability to Java. In *Proceedings of the 20th Annual ACM SIGPLAN Conference on Object-Oriented Programming, Systems, Languages, and Applications* (pp. 211–230).

UN News Center. *Corruption leading to unequal access.*

Valenzuela, J. (2015). *Bitcoin remittances to Mexico see huge potential.* Cointelegraph. https://cointelegraph.com/news/bitcoin-remittances-to-mexico-see-huge-potential

Vanham, P. (2018). *Blockchain could enable $1 trillion in trade, mostly for SMEs and emerging markets.* World Economic Forum. https://www.weforum.org/press/2018/09/blockchain-could-enable-1-trillion-in-trade-mostly-for-smes-and-emerging-markets/

Vidrih, M. (2018, May 22). *The blockchain international standardization organization: China will dominate.* Good Audience. https://blog.goodaudience.com/the-blockchain-international-standardization-organization-china-will-dominate-b7d423904bfb

Walk Free Foundation, The 2018 global slavery index https://downloads.globalslaveryindex.org/ephemeral/GSI-2018_FNL_190828_CO_DIGITAL_P-1596333264.pdf

Wang, C. (2018, December 13). *Tencent integrates blockchain e-invoicing with WeChat.* 8BTC. https://news.8btc.com/tencent-integrates-blockchain-e-invoicing-with-wechat

Wass, S. (2018, January 24). *Major banks and traders test blockchain platform for commodity trade*. Fintech. https://www.gtreview.com/news/fintech/major-banks-and-traders-test-blockchain-platform-for-commodity-trade/

Wass, S. (2019, June 25). *komgo releases two new blockchain-based products, reveals future plans*. Fintech. https://www.gtreview.com/news/fintech/komgo-releases-two-new-blockchain-based-products-reveals-future-plans/

Willis Towers Watson. (2016, June). *Want to get an insurer's attention? Just say blockchain*. https://www.the-digital-insurer.com/wp-content/uploads/2016/12/793-want-to-get-an-insurers-attention-just-say-blockchain-wtw.pdf

White, O., Madgavkar, A., Manyika, J., Mahajan, D., Bughin, J., McCarthy, M., & Sperling, O. (2019). *Digital identification: A key to inclusive growth*. McKinsey.com. www.mckinsey.com/business-functions/digital-mckinsey/our-insights/Digital-identification-A-key-to-inclusive-growth?cid=other-eml-alt-mgi-mck&hlkid=ecd8822bafc44de78b1f2670c2979652&hctky=2259579&hdpid=0945f28e-3aa8-4f73-a111-6ba86e377b51

Williamson, O. E. (1985). *The economic institutions of capitalism*. The Free Press.

Wong, J. I., & Huang, Z. (2016, October 21). *China's tech giants will shape international blockchain standards, with Beijing's backing*. Quartz. https://qz.com/813248/chinas-tech-giants-will-shape-international-blockchain-standards-with-beijings-backing/

World Bank. (2020). *Doing business measuring business regulations*. https://www.doingbusiness.org/en/data/exploretopics/enforcing-contracts

Yaga, D., Mell, P., Roby, N., & Scarfone, K. (2018, October) *Blockchain technology overview* (National Institute of Standards and Technology Internal Report (NISTIR) 8202).

Zwitter, A., & Boisse-Despiaux, M. (2018). Blockchain for humanitarian action and development aid. *Journal of International Humanitarian Action, 3*(1), 16. https://doi.org/10.1186/s41018-018-0044-5

CHAPTER 2

Political and Administrative Consequences

Abstract This chapter examines blockchain's roles in improving account-ability, responsibility, and efficiency in the public sector in developing countries. It evaluates various mechanisms and features of blockchain that can help prevent fraudulent and corrupt practices. Specifically, using practical applications and real-world examples that have been or are being developed, it analyzes how blockchain can help to fight corruption in public procurement. It will discuss how informal account-ability is among the key mechanisms facilitated by blockchain that can delegate some of the functions to other stakeholders in order to fight corruption in public procurement. It also advances our understanding of how blockchain can enhance efficiency in customs administration and facilitate trade. It gives special consideration to blockchain's use in land registries to reduce corruption and increase efficiency in land-related transactions. It also gives an overview of blockchain's roles in facilitating rapid response during natural disasters and political crises.

Keywords Accountability · Bribes · Bureaucrats · Corrupt practices · Customs administration · Responsibility

© The Author(s), under exclusive license to Springer Nature   29
Switzerland AG 2023
N. Kshetri, *Blockchain in the Global South*,
https://doi.org/10.1007/978-3-031-33944-8_2

2.1   INTRODUCTION

Different forms of unethical and corrupt practices linked to the lack of responsibility, accountability, integrity, and transparency are of concern in the developing world (Hope & Chikulo, 2000). In many GS economies, public sector corruption acts as a hindrance to the proper functioning of the market system, which lowers economic growth (Osterfeld, 1992). In Africa, bureaucrats, civil servants are reported to illegally increase their income by providing services to interest groups that want favors from the government (Mbaku, 1996). The annual costs of corruption worldwide are estimated at US$3.6 trillion (Johnson, 2018), about $1 trillion of which is paid in bribes (Hartson, 2016). In the healthcare sector alone, about $500 billion is lost to corruption annually (Bruckner, 2019). In many cases, corruption also leads to the loss of lives (Brende & Gomez Pensado, 2020). Especially public procurement, which involves the purchase of goods and services for government departments and state-owned enterprises has been the main area of corruption in these economies (Mawenya, 2008).

Prior research has also noted that public bureaucracies in developing as well as developed countries function as monopolies that lack effective mechanisms to protect against inefficiency (Landau, 1991). Bureaucrats also tend to prefer an inefficient state structure, which allows them to gain more rents compared to an efficient one (Acemoglu et al., 2011).

Blockchain has been touted as a transparency-enhancing tool with the potential to address the various challenges noted above. Some politicians have advocated using blockchain for fighting economic crimes such as corruption and embezzlement. Solomon Adaelu, a member of the House in Nigeria argued that blockchain can help eradicate corrupt practices in both the public and private sectors of the country (Avan-Nomayo, 2019).

There are various mechanisms and features of cryptocurrency and blockchain that can help reduce fraudulent and corrupt practices. Due to easily accessible information and faster crosschecks, they reduce the costs of enforcement. They also help supervise a project's implementation and monitor the efficiency and effectiveness of spending in the project (Aldaz-Carroll & Aldaz-Carrol, 2018). Blockchain can also reduce transaction costs and increase equity, performance, and efficiency in law enforcement and the court system. For instance, police corruption has been a challenge in developing countries. In South Africa, such corruption has led to a seemingly justifiable low level of public trust in South African Police Service (SAPS) (Newham & Faull, 2011). One way to produce and sustain trust in the police would be to use blockchain. For instance, police body camera, dash camera, and police report data can be hashed and stored in public blockchains to create increased transparency between police and the public. Such data can also empower the legal community since police cannot tamper with evidence (Greenfield, 2017).

Some real-world blockchain applications to enhance efficiency in court system and other administrative machinery have been built. For instance, China's courts are applying blockchain and artificial intelligence to settle millions of legal cases (xinhuanet.com, 2019). The Chinese government has also announced a plan to start requiring individuals, and companies to submit their applications for tax reimbursement via a blockchain system (Mazor, 2018).

What makes blockchain interesting is its potential to fight problems such as corruption, inefficiency, and mismanagement, which are more serious problems in developing countries than in developed countries. Prior researchers have also argued that blockchain has a much higher value proposition for the developing world than for the developed world (Kshetri & Voas, 2018). The aim of this study is to provide further insights into this issue through an examination of the role of blockchain in fighting corruption and inefficiency in public administration.

2.2   Key Political and Administrative Challenges Facing the GS and the Roles of ICTs

Accountability, Responsibility and Efficiency

Accountability is viewed as a complex and "chameleon-like" term in the public administration literature (Mulgan, 2000). The most agreed-upon definition of accountability is arguably associated with the processes that *require individuals* "to account" for their actions *to* an authority (Jones, 1992, p. 73).

By responsibility, we mean accountability in some specific way to some type of constituency. Responsibility also means a sense of moral obligation and integrity. Prior research has emphasized the importance of administrative responsibility in a democratic government and noted that administrative responsibility may contribute to administrative efficiency in the long run (Finer, 1941).

A concern is that responsibilities are not being adequately fulfilled by governments worldwide. For instance, Latin America's national governments have been accused of abandoning their responsibility to fight poverty (Fox, 1994).

Due to increasing use of corporate social responsibility (CSR) as an instrument for global governance, functions that were previously performed by the state are being transferred to private companies or NGOs. These trends have blurred the lines of responsibility and accountability of private and public actors (Campbell, 2012). A number of factors such as personal values and national norms concerning the relationship between individuals and the state and responsibility of government and complex management issues will influence the decision regarding whether certain functions should be performed by the states or can be delegated to a non-governmental organization (Cohen, 2001).

Informal accountability provides another mechanism to delegate some of the functions. Prior researchers have noted that compared to formal aspects of accountability in contract relationships, relatively less attention has been focused on informal accountability such as interorganizational and interpersonal behaviors. Various elements of informal accountability such as those related to social norms and individual actions and behaviors can play significant complementary roles to formal accountability, which can enhance network performance (Romzek et al. 2012).

A point that is worth noting is that responsibility is a prerequisite to efficiency (Finer, 1941). A government is also entrusted with the responsibility to deliver administrative services efficiently. For many developing countries, inefficient government administration can be a key barrier to economic development. For instance, inefficient government bureaucracy makes doing business a difficult and expensive process in African economies (Hartzenberg, 2011). Countries are losing about one-third of the outputs they could deliver due to inefficiency (Afonso et al., 2010).

Corruption

Corruption, which is tightly linked to the lack of accountability, responsibility, and efficiency, is a major problem in the public sector in developing countries. According to a 2011 report of the UN Food and Agriculture Organization (FAO) and Transparency International, in over 61 countries, weak governance led to corruption in land occupancy and administration. Corruption varied from small-scale bribes to the abuse of government power *at* the national, state, and local levels (UN News Center, 2011).

In a regulated economy, Osterfeld (1992) has identified two types of corruption:

a) Expansive corruption involves activities that improve the market's competitiveness and flexibility. Some examples include the private sector's paying bribes to judges, politicians, and bureaucrats. Such activities arguably mitigate the harmful effects that excessive government regulations can have and improve economic participation.

b) Restrictive corruption restricts opportunities for exchanges that are productive and socially beneficial. This type of corruption involves public resources' illegal appropriation for private use. It leads to a redistribution of income and wealth in favor of certain individuals or groups. Most public-sector corruption is of this type.

Klitgaard has identified three necessary conditions of corruption: (1) there is the opportunity for an economic rent, (2) agents have discretionary powers, and (3) there is a lack of accountability (Klitgaard, 1988). The above necessary conditions are perfectly satisfied in public procurement. Corruption in public procurement involves the "misuse of public power in procurement for private benefits" (Piga, 2011). The involved agents are in a position to generate economic rent by artificially inflating the contract's value with ex ante manipulation such as price fixing or ex post manipulation, such as contract modification. Government agencies can use their discretionary power to get around the ideal procurement process. Cost overrun-related issues often go unquestioned, which lead to a lack of accountability (Hudon & Garzón, 2016).

Governmental procurements involve complex and opaque tendering processes, close interactions between public officials and businesses, and a large number of stakeholders (Barrera et al., 2019). Due to these factors, government tendering and public procurement are high-risk areas for corruption (Rustiarini et al., 2019), especially in developing countries (OECD, 2016).

Government procurement systems in GS countries are closely controlled by powerful government actors and politicians, who often misuse their power. In order to reform, government procurement systems, issues of accountability and transparency need to be addressed and a professional workforce must be developed (Raymond, 2008).

The Roles of ICTs

The "Panoptic vision" views ICTs as a key management control mechanism (Heeks, 1998). To put things in perspective, when the Internet was new, there was a big hope that it would promote corporate and government transparency. Some analysts predicted that the Internet would make it impossible for governments and corporations to hide anything (Tapscott & Ticoll, 2003). However, soon it turned out to be a naïve view. While ICTs sometimes can detect and help fight corruption, they have no effect on some types of corrupt practices and may provide new corruption opportunities in some cases (Heeks, 1998).

ICTs also have efficiency-enhancing and trust-building effects on public institutions. E-government can improve the administrative efficiency of public institutions, encourage democratic governance, and build trust between citizens/private sector and governments (Verkijika, & De Wet, 2018).

Governments have started to implement new ICTs to improve the integrity and efficiency of their procurement processes. E-procurement systems have various integrity risks such as insider frauds (e.g., misuse of firewalls to restrict bid submission, document changes by system admins, and leakage of confidential information), Outsider attacks (e.g., hacking) and other design flaws (e.g., flawed bid sealing/encryption for guaranteed confidentiality, digital signatures not supported) (Kohli, 2012).

An application of ICTs in government administration that affects wider society is land records' digitization. Mixed results have been obtained in studies assessing the impact of the digitization of land records. For instance, various benefits of the Electronic Document Management System (EDMS) implemented in Lagos, Nigeria were identified. Land-related files were stored centrally, which facilitated the search and retrieval of documents. Waiting time to obtain land-related information was reduced and administrative efficiency improved. Overall, it led to an increase in public confidence (Thontteh & Omirin, 2015).

As another example, one can look at the Bhoomi program' in India's Karnataka state (Benjamin et al., 2007). A computerized database of 20 million land records that belonged to 6.7 million farmers was generated. In the pre-Bhoomi era, there were many different land ownership forms due to various socio-legal processes that underpinned land claims. Bhoomi unified these heterogeneous land tenure forms (Benjamin & Raman, 2011). In the pre-Bhoomi system, land-related corruption were related to ambiguity in land records, in which the process was mostly controlled by local-level "street bureaucrats" (Benjamin & Raman, 2011). To some extent, digitization helped overcome such problems.

However, several limitations and drawbacks of digitization initiatives such as Bhoomi and the EDMS have been noted. For instance, the EDMS showed no improvement in resolving boundary disputes. Moreover, the EDMS system did not lead to an increase in the number of applications processed. No additional revenue was generated due to the implementation of the system (Benjamin & Raman, 2011).

An analysis of the Bhoomi system indicated that the digitization of land records does not necessarily improve farmers' welfare. The record also included a history of cropping patterns of the previous 12 seasons. This was arguably the digitization of land cadastrals at the largest scale carried out in a developing country. The Bhoomi system increased corruption and bribes as well as time taken for land transactions (Benjamin et al., 2007). The digitization was carried out by the centralization of land records and the management moved away from villages to taluk offices at the district level.

Before Bhoomi, obtaining a copy of a Record of Rights, Tenancy and Crop Information (RTC), or mutation—the transfer of rights from one owner to another—normally took 2–3 days. The Village Accountants (VAs) made such documents available efficiently. After Bhoomi's implementation, a mutation took as long as four months and farmers needed to depend on agents to get their work done. In many cases, when farmers visited the taluk office losing their work, they found that computers were down or there was no electricity.

It was reported that large and middle-level farmers exploited the centralization and computerization of land records *to their benefits. These groups allegedly used* smaller farmers' survey numbers to access government schemes and benefits such as subsidies provided to small farmers to buy seeds, fertilizers, and pesticides. Bhoomi *also helped powerful economic groups to acquire lands at prime* locations *at a low* cost from small farmers, who had used such lands for cultivation. In some cases, the farmers had received such lands under various so-called "inam" schemes, which restricted them to sell. Economically and politically connected real estate agents contacted several small landholders in an area, nego-tiated, and *converted the land to commercial* use. To do so, the urban economic agents obtained No Objection Certificate (NOC) from the village panchayat and used the Bhoomi system to issue titles for their clients. Large land developers had connections with politicians and senior administrators. In such cases, due to the involvement of *powerful and influential interests*, the local institutions such as the VA office or the Bhoomi kiosk had no power to influence the process or make a decision.

Bhoomi shows that the process of digitization and related digital access to land titles may shift power and wealth to those with the financial resources and skills. These groups are in a position to use this informa-tion to promote their self-interest and harm the interests of disadvantaged groups such as small farmers (Benjamin et al., 2007).

2.3   BLOCKCHAIN IN GOVERNMENT ADMINISTRATIVE SYSTEMS IN THE GS

Among major technologies, blockchain holds strong promise to fight against corruption and inefficiencies. Blockchain transactions are immutable. Once they are entered into the ledger, they cannot be modi-fied or deleted. Cryptography-based authentication means that relevant participants have a digital identity on every transaction. These features make manipulation next to impossible. Governments are increasingly adopting blockchain to ensure responsibility and accountability, enhance efficiency, and fight corruption.

National-level Strategies, Policies, and Measures to Develop Blockchain Solutions

Before preceding further, it is important to note that national-level strategies and policies are in place, or under development in many devel-oping countries to improve political and administrative systems utilizing blockchain. In 2018, Kenya's then president Uhuru Kenyatta launched an eleven-member blockchain and artificial intelligence (AI) task force to create a 15-year roadmap that can help reduce political corruption and red tape (De Meijer, 2018). Likewise, in 2018, Blockchain Associ-ation of Uganda was established to drive standards for blockchain across industries. It also aims to make blockchain-related resources available to government agencies (Global Banking & Finance Review, 2018).

The UAE is also promoting blockchain. *The Dubai Future Foundation, which was* launched by the Dubai government to explore emerging technologies established Global Blockchain Council. High-profile players such as Etisalat, Emirates NBD, du, Dubai Multi Commodities Centre, IBM, and Microsoft have joined the Council (Young, 2021). Dubai's plan is to secure all official documents with blockchain technology in order to simplify interactions between government agencies and citizens. It also aims to simplify businesses' interaction with local authorities (Maltaverne, 2021). Dubai Customs is developing a collaborative blockchain-based Cross Border e-Commerce Platform (tradearabia.com, 2021).

Some governments are developing technological solutions that can help fight corruption and increase administrative efficiency. In order to reduce corruption and fraud, in 2017, Brazil's state-run technology company Serpro launched a blockchain platform and held roadshows to introduce the technology across the country (Medes, 2018). In November 2020, Brazil's federal government published Decree No. 10.550, in the Federal Official Gazette, which aimed "to adapt it to the recent technological advances in foreign trade systems." The Decree included the regulation of the use of electronic signature and blockchain (Mota, 2020).

In 2018, Serpro launched a blockchain platform to regulate land titles in the country (Mendes, 2018). A goal of the platform is to reduce corruption. Every action taken in the system is recorded, which means that government officials cannot delete files without being noticed by others (nasdaq.com, 2018).

Fighting Corruption in Public Procurement

From the corruption standpoint, public procurement has been one area that has received a great deal of attention. Governments world-wide spend US$9.5 trillion annually on public procurement, which is about 15% of the global GDP (Davidson et al., 2020). About 10–30% of a public contract's value is lost due to corruption (OECD, 2016). According to the Organization for Economic Co-operation and Development (OECD), 57% of foreign bribery cases occurred in order to obtain public procurement contracts (OECD, 2014).

A high-profile example of firm engaged in corruption in public contracts in the Brazilian construction firm Odebrecht S.A. Between 2001 and 2016, Odebrecht S.A. engaged in illegal campaign financing and large-scale bribery of about US$800 million in order to secure over 100 contracts across 12 countries, which included Colombia (https://www.justice.gov/opa/press-release/file/919911/download) (Barrera et al., 2019).

Blockchain can limit problems such as manipulation in public procurement processes (transparency.org, 2018). The shared and immutable records cannot be censored or altered by government agencies. Records of bids and public comments cannot be deleted and a vendor cannot be *denied from bidding*. Bids or tender offers cannot be altered once they are submitted (Davidson et al., 2020). These features make blockchain an effective tool to fight corruption in procurement. Such benefits are especially likely to accrue from permissionless blockchains like Ethereum (Allison, 2020).

Some international agencies have also advocated for the use of blockchain in fighting corruption in public procurement. The United Nations Office on Drugs and Crime (UNODC) has suggested the Kenyan government use blockchain to fight economic crimes. Government officials in Kenya allegedly manipulate procurement procedures and systems to inflate costs for their own gains. The country's highest offices including the vice president have been connected to scandals (Kaaru, 2020). According to Kenya's Auditor General, the country loses US$10 billion annually to corruption (Isaac, 2019).

Some governments have been serious about implementing blockchain in public procurement (In Focus 2.1). Peru's government announced a partnership with blockchain startup Stamping.io to create a blockchain-based contract-procurement system. The goal is to create a verification system for government contracts that is resistant to data manipulation and other frauds. It will register purchase orders from the government agency to regulate electronic purchases Peru Compras. It uses LAC-Chain, the Inter-American Development Bank's private blockchain whose nodes are managed by the IDB (perureports.com, 2019).

In Focus 2.1: Blockchain in Colombia's School Meal Procurement
In Colombia, many corrupt practices have been reported in school meal procurement. *Various scandals have* arisen in this program in recent years. The Colombian newspaper *El Tiempo* reported that chicken breasts were sold to schools at $12 (COL$40,000 (Colombian peso)), which was

about four times the price of local supermarkets (eltiempo.com, 2017). In some cases, goods that are purchased are not delivered. The former mayor of the port city on the Caribbean coast Cartagena was charged for illegally contracting a deal of over COL$23 million (about $7,000). Out of 2.6 million loaves of bread that were bought, one million were never delivered to schools. Public figures and officials and a small number of food contractors have been involved in procurement frauds (Sarralde Duque, 2017).

In an attempt to address corrupt practices in public procurement such as the one discussed above, the World Economic Forum has teamed up with the Inter-American Development Bank (IDB) and the Office of the Inspector General of Colombia (Procuraduría General de Colombia) to investigate, design and trial the use of blockchain for public procurement activities. A software proof of concept focused on a school meals program (Programa de Alimentación Escolar) which targets young people from low-income households was developed. A public blockchain procurement system was used to track the process of supplier selection in this program in the city of Medellín. Ethereum blockchain was used (Hall, 2020).

The WEF's initial Blockchain project in Colombia focused on the contractor selection phase. The goal is to improve transparency, fairness, and competitiveness in the bidding process. A tenderer publicly commits to contract terms and selection criteria prior to eliciting bids. In this way, risks such as tailoring selection criteria *after the RFP is published* to favor specific contractors are eliminated. For vendors that are competing, a blockchain-based solution's permanent and tamper-proof bid records can ensure that a firm cannot alter submitted bids after learning new information about competing bids. An additional benefit is that by increasing the perception of fairness, blockchain-led transparency can attract more vendors to the procurement process. A set of clearly defined selection criteria would increase the possibility that an outsider can win. During the auction and vendor evaluation processes, actions and decisions are automatically recorded. These records are permanent and publicly viewable, which increases auditability. It is also possible to include a user interface, which can be used by the public to monitor actions and decisions so that risks can be flagged in real-time. These enable monitoring authorities such as the Inspector General's Office to investigate potential corrupt activity even before an auction concludes.

Blockchain-based solutions can also be used to monitor the chosen contractor's performance. For instance, information regarding actual deliveries can be made available to key stakeholders such as parents, teachers, enforcement officials, and the press. Their participants can be used to report meal deliveries and quality in real time. By improving observation in the delivery process and allowing stakeholders to monitor and engage, accountability of contractors can be improved (Barrera et al., 2019). By allowing the participation of diverse groups, such systems can promote informal accountability, which, as prior research has shown (Romzek et al., 2012), can improve the performance of public procurement.

Enhancement of Efficiency in Customs Administration and Facilitation of Trades

Blockchain holds promise in driving efficiency in customs administration and facilitate trade. Every shipment involves the exchange of around 200 documents and 300 people (Wood, 2020). Efficiency is greatly enhanced. A number of blockchain-based solutions have been launched for such purposes.

Inter-American Development Bank's (IDB) CADENA

The World Customs Organization has developed a framework to identify secured and trusted actors, known as, the Authorized Economic Operators (AEO). As of 2019, about 80 countries had compiled lists of entities and certified that they meet AEO standards. Customs can focus their resources on non-certified actors (Corcuera-Santamaria & Leonor Moreno, 2019). The system's benefits can be maximized if customs administrations share their lists of AEOs with their counterpart agencies in other countries. Without sharing such information, exporters can get expedited treatment only in their county but not in the destination. As of 2019, 60 mutual recognition arrangements (MRAs), which involve sharing of lists, had been signed and 40 more were being negotiated. Some of the MRAs are bilateral, while others are multilateral.

The customs administrations of Pacific Alliance (https://alianzapacifico.net/en/what-is-the-pacific-alliance/) member countries Colombia, Chile, Mexico, and Peru had bilateral MRAs. The lists of AEOs change. New entities can be added and some can be removed. Customs officers send Excel files containing the data of their AEOs by emails to their counterparts usually every month. A challenge is that email systems are not secure. When entities are added or removed from the AEO, there is often a delay in communicating the list to counterpart customs agencies.

To address these challenges, the Integration and Trade Sector of the Inter-American Development Bank (IDB) teamed up with AEO program officers and IT specialists from Mexico, Peru, Chile, and Costa Rica and Microsoft to develop a blockchain-based application called CADENA. CADENA works on real time. It increases transparency and trust.

BConnect

BConnect is a blockchain network developed by Serpro. Firms can use BConnect to ensure the authenticity of customs information shared between Mercosur countries. Since October 2020, the network has been in operation to exchange data between Argentina, Brazil, Paraguay, and Uruguay. BConnect makes it possible to share registration information of companies certified by the Federal Revenue as Authorized Economic Operator (*OAS*). Such companies enjoy benefits, such as the facilitation of customs procedures in Brazil and abroad.

BConnect ensures the integrity of information shared between countries and verifies the identity of those feeding information onto the platform. The *Receita Federal* commissioned Serpo to develop BConnect. If a Brazilian company wants to export to a company in Uruguay, the fact that both of them are registered on BConnect makes vetting process easier for Uruguayan (López, 2019).

BConnect uses Hyperledger (Isabelle Guzzo, 2020). Hyperledger Fabric's advantages over other open blockchains include data protection and consistency, use of permissions to ensure control of members and access rights, and confidential transactions (latam.portalerp.com, 2020).

Land Registry

Some scholars have suggested that land titling systems could be blockchain's "low hanging fruit" applications (Swan, 2017). In general, blockchain can arguably be applied to property registry in an inexpensive way (Dwyer, 2016) and recording any property transactions in blockchain could offer several benefits (Gabison, 2016).

In most developing countries, value, ownership, and other details of lands are in paper-based cadastres, which are mostly incomplete. These pose significant challenges in digitizing and updating land records in order to accurately reflect ownership of property (Kriticos, 2019). It is important to tackle the widespread land governance challenges in order to increase the scalability of the implementation of blockchain projects. A major challenge is to convert analogue land registers and analogue *cadastre to digital* land databases. Buenos Aires cleaned up the analogue system and digitized the registry. In such cases, blockchain can have major advantages by ensuring immutability.

Land registry is becoming an increasingly popular application of blockchain (Manski, 2017). Various benefits of blockchain's use in developing a national system for property management have been suggested (Lemieux, 2017). Both actual and proposed implementations of blockchain for land registry systems in Honduras (Lemieux, 2016), Ghana (Kshetri, 2017b), Georgia, India, and other countries (Kshetri & Voas, 2018) have been given as examples to illustrate the significance of blockchain.

The U.S.-based platform for real estate registration, Bitland announced the introduction of a blockchain-based land registry system in Ghana, where 78% of land is unregistered (Ogundeji, 2016). There is a long backlog of land-dispute cases in Ghanaian courts (Jones, 2016). About 90% of the land is undocumented or unregistered in rural Africa. Bitland records transactions securely with GPS coordinates, written descriptions, and satellite photos. The process is expected to guarantee property rights and reduce corrupt practices. As of mid-2016, 24 communities in Ghana had expressed interest in the project (Ogundeji, 2016).

Bitcoin company BitFury and the Georgian government signed a deal to develop a system for registering land titles using the blockchain (Higgins, 2016a). The Peruvian economist Hernando de Soto would assist in the development of the platform. In order to buy or sell land in Georgia, currently the buyer and the seller go to a public registry house. They are required to pay between $50 and $200, which depends on the speed with which they want the transaction to be notarized. The pilot project will move this process onto the blockchain. The costs for the buyer and the seller are expected to be in $0.05-$0.10 range (Higgins, 2016b). With blockchain-based land titling project, the government of Georgia aims to enable landowners to borrow against their lands and engage in entrepreneurial activities (Manski, 2017).

In India's Andhra Pradesh State, the TDP government headed by then Chief Minister Chandrababu Naidu announced plans to use blockchain for land registry. It was the earliest Indian state to do so. A typical land record in blockchain includes 58 attributes (Sai Baba, 2020). They include static attributes that describe the property, such as unique ID, plot code, geo-coordinates (latitude and longitude), survey number, boundary Information (e.g., information about neighboring plots, location in relation to roads or other landmarks) classification of land as well as dynamic attributes that are subject to change such as owner (e.g., Aadhaar number) and mortgage information, right of first refusal (*ROFR) and* litigation status. Events *such as m*utation, court case filing, stay issued by the court, sale, approval of buildings, conversion of lands (e.g., from agricultural to commercial), mortgage, and the owner's death are also recorded. The system also provides flexibility to add new attributes if such needs arise in the future (Sai Baba, 2020).

The Andhra Pradesh land project uses permissioned blockchains. The nodes in Andhra Pradesh's land records include the Revenue Department, Chief Commissioner of Land Administration, and other officials. A key advantage of private or permissioned blockchains is the higher speed to process transaction data compared to permissionless blockchains as measured in transactions per second (TPS). A limitation of private blockchains such as the one used in AP is that individual landowners cannot access their records on the blockchain. However, QR code-based property ownership certificates, which provide details about all transactions, are issued to the owners (Sai Baba, 2020).

Prior research has suggested that in the absence of appropriate measures digitization of land records may lead to manipulation of the land market process by powerful actors (Benjamin et al., 2007). A blockchain-based system in which many agencies act as nodes or validators of transactions could serve as a *check and balance for* one another to assure that no agency can manipulate the system without being noticed by others. Blockchain-based land recording systems may use various types of consensus algorithms such as proof of work and proof of stake consensus for validating transactions, in which all or some nodes verify transactions such as a change in the landowner's name for a plot (Vos et al., 2017). Following this, a new block of data involving land transactions is added to the ledger.

As noted above, Andhra Pradesh's blockchain-based land recording system relies on a small number of nodes to validate land-related transactions. In general, a blockchain with only a small number of data validators is considered to be less trustworthy compared to another blockchain with a large number of such validators (Jagati, 2020). Despite this situation, while bribes are not impossible, such activities face considerable challenges in blockchain-based records. In the centralized model, *powerful and influential actors can pressure the individuals who* manage the centralized database such as the Bhoomi system *to* change records. In a blockchain-based system, records cannot be tampered with without being noticed by other nodes. Moreover, in the Andhra Pradesh system, if any node makes an attempt to change the record, the landowner will receive a text message (ENS, 2019).

Currently, after a land transaction is finalized, the officer in charge of the collection of land revenues (tehsildar) needs to submit a land demarcation in order to register the deed. The process takes between one to three months. Bribes are often paid to prepare documents (Ramnani, 2018). With blockchain, properties can be transferred in a day without paying bribes.

Blockchain can also lead to important cost-saving opportunities. Before the implementation of blockchain, farmers needed to pay at least US$68 to prepare registration papers in the Andhra Pradesh state. Now, they can get system-generated digital documents for free. The digital document with a QR code can be sent directly to the land registrar for transactions(Bhattacharya, 2018).

Rapid Response During Natural Disasters and Political Crises

During natural disasters and political crises, blockchain-based solutions can help rapidly respond and disburse cash assistance to affected families. For instance, ethereum blockchain, which can be used to create a new cryptocurrency, is being used for such purpose. Additional information can be included in the blocks. Smart contract features can be used. By setting up the platform, governments or development institutions can allocate cryptocurrency to the activities to be performed. The entities assigned to conduct the activities take the cryptocurrency and spend it in goods and services. Verification is performed in the Ethereum platform. Access to the cryptocurrency can be provided using software wallets, which do not require a bank account. This is a key advantage in developing countries with low financial access. The holders of cryptocurrency can convert it into fiat currency at an exchange market. it can be done in a secondary exchange market or a primary exchange market run by the development institution or the government. They can also use it as currency if vendors accept it (Aldaz-Carroll, & Aldaz-Carroll, 2018).

To take an example, the World Food Program's (WFP) Innovation Accelerator started "Building Blocks" pilot in early 2017. In the first stage, food and cash assistance were provided to needy families in Pakistan's Sindh province. Starting in May 2017, the WFP started distributing food vouchers in Jordan's refugee camps by delivering cryptographically unique coupons to participating supermarkets. Supermarket cashiers are equipped with iris scanners to identify the beneficiaries and settle payments. UN databases verify biometric data about refugees. Building Blocks' ledger records the transactions on a private version of the Ethereum blockchain: the Parity Ethereum. No banks are involved and beneficiaries thus receive goods directly from the merchants.

Individuals are given encrypted IDs or code numbers and they are not required to reveal true identities. The Parity Ethereum used in the system employs four nodes to validate transactions (Stanley, 2021). This means that transactions cannot be seen by actors that are not a part of the authorized peer nodes. An additional benefit is that the cryptocurrency mining process is not needed to validate the transactions. This feature removes a key bottleneck to the processing speed and transaction capacity (Wong, 2017). The system is designed to scale.

The WFP reported that by October 2017, it had distributed US$1.4 million in food vouchers to 10,500 Syrian refugees in Jordan (Kshetri & Voas, 2018). As of early 2019, 1.1 million cryptocurrency transactions transferred more than US$ 23.5 million to refugees. As of late 2020, Building Blocks assisted over 400,000 refugees around the world, including more than 100,000 Syrian refugees living in camps in Jordan (Ledger Insights, 2020b).

When the applications reach a more advanced development stage in the future, more benefits can be realized. For instance, the WFP expects that refugees may be able to access their funds by controlling their cryptographic keys. This would also allow them to incorporate and integrate personal data from diverse sources. For instance, their medical records could be with the World Health Organization (WHO), academic credentials with the United Nations Children's Fund (UNICEF), and nutritional data with the WFP (Wong, 2017). In this way, they can build their economic identity.

Similar benefits can be achieved in the systems used to distribute donations and aid. For instance, the WFP expects that blockchain-based solutions would reduce its overhead costs from 3.5% to less than 1% (Kshetri & Voas, 2018). What is even more important is that an estimated 30% of development funds do not reach the intended recipients due to problems such as third-party theft and mismanagement (Paynter, 2017). Blockchain holds great potential and promises to reduce such practices.

2.4   Chapter Summary and Conclusion

Blockchain can allow government agencies to fulfill their mission and responsibility in an effective manner by managing scarce resources in a responsible manner. Blockchain systems can be developed to promote transparency and accountability of goals in public administration. Blockchain systems can help the prevention of irresponsible behavior by government officials. Blockchain-based evidence can help promote transparency, accountability, and effectiveness and increase the perception of justice and fairness in law enforcement. Even if there is no corruption involved, blockchain-based solutions can play a key role in the pursuit of efficiency in public administration.

Some encouraging initiatives to deploy blockchain to fight poverty are ongoing. Blockchain projects such as land registry can improve openness, transparency and accountability as well as efficiency. Such a system can also facilitate easy access to capital by using the property as collateral to secure financing, which can stimulate entrepreneurial activities.

Some limitations and challenges of blockchain systems need to be pointed out. For instance, blockchain can present a challenge to government control over the population. The Nigerian example indicates that young and tech-savvy Nigerians have been able to use cryptocurrencies to navigate a complicated and restrictive banking and monetary system and circumvent political controls.

Finally, while blockchain systems are secure, their data—like other databases—are only as accurate as what is entered. In the land ownership example, blockchain can increase the transparency of land ownership records and make it difficult or impossible for corrupt officials to alter land registries after the *records are on the blockchain. Nonetheless,* blockchain cannot address *corruption in decisions* about how land is registered in the ledger.

REFERENCES

Acemoglu, D. Ticchi, D., & Vindigni, A. (2011). Emergence and persistence of inefficient states. *Journal of European Economic Association, 9*(2), 177–208. https://academic.oup.com/jeea/article-abstract/9/2/177/2298409

Afonso, A., Schuknecht, L., & Tanzi, V. (2010). Public sector efficiency: Evidence for new EU member states and emerging markets. *Applied Economics, 42*(17), 2147–2164. https://www.tandfonline.com/doi/full/10.1080/00036840701765460?casa_token=Sa8ms3oqHOgAAAAA%3AN0lZD4DdwxU2F86A3hNJwxdDgfL67O7YdEnJSq23I7a0ULoJd1rz4bR9DTXpxt4x0lPyXh5FjtC0pw

Ahmed Mahboob Musabih Exceptional achievements by Dubai Customs in 2020. (2021). *Exceptional achievements by Dubai Customs in 2020.* http://tradearabia.com/news/MISC_377138.html.

Akshatha, M. (2018). Karnataka's famed land record database Bhoomi faces another security breach, *The Economic Times.* https://tech.economictimes.indiatimes.com/news/corporate/karnatakas-famed-land-record-database-bhoomi-faces-another-security-breach/65748534

Aldaz-Carroll, E., & Aldaz-Carrol, E. (2018). *Can cryptocurrencies and blockchain help fight corruption?* Brookings. https://www.brookings.edu/blog/future-development/2018/02/01/can-cryptocurrencies-and-blockchain-help-fight-corruption/

Allison, I. (2020). *Colombian government and WEF Weigh Public Ethereum in bid to fight corruption.* Coindesk. https://www.coindesk.com/colombian-government-and-wef-weigh-public-ethereum-in-bid-to-fight-corruption

Antonio Lanz, J. (2019). *Peru sets its eyes on blockchain to fight government corruption.* Decrypt. https://decrypt.co/6893/peru-blockchain-government-corruption

Avan-Nomayo, O. (2019). *Africa using blockchain to drive change, part one: Nigeria and Kenya.* CoinTelegraph. https://cointelegraph.com/news/africa-using-blockchain-to-drive-change-nigeria-and-kenya-part-one

Baba, S. (2020). Aphrdi—Andhra Pradesh, HRDI. https://www.aphrdi.ap.gov.in/.

Barrera, C., Hurder, S., & Lannquist, A. (2019). Here's how blockchain could stop corrupt officials from stealing school lunches. *World Economic Forum.* https://www.weforum.org/agenda/2019/05/heres-how-blockchain-stopped-corrupt-officials-stealing-school-dinners/

Benjamin, S. R., Bhuvaneswari, & Rajan, P. (2007). Bhoomi: 'E–governance', or, an anti–politics machine necessary to globalize Bangalore? *CASUM–m Working Paper.* https://www.semanticscholar.org/paper/Bhoomi-%3A-%E2%80%98-E-Governance-%E2%80%99-%2C-Or-%2C-An-Anti-Politics-Benjamin-Bhuvaneswari/f07edf3436b77a7888558ba95e37d3951875703e

Benjamin, S., & Raman, B. (2011). Illegible claims, legal titles, and the worlding of Bangalore. *Dans Revue Tiers Monde, 2*(206), 37–54.

Bhattacharya. (2018). *Blockchain is helping build a new Indian city, but it's no cure for corruption.* Quartz. https://qz.com/india/1325423/indias-andhra-state-is-using-blockchain-to-build-capital-amaravati/.

Boeding, K., & McConkie, R. (2021). *3 potential benefits of blockchain for government.* Booz Allen Hamilton. https://www.boozallen.com/s/insight/blog/3-potential-benefits-of-government-blockchain.html

Brende, B., & Gomez Pensado, G. P. (2020). 3 ways to fight corruption and restore trust in leadership. *World Economic Forum.* https://www.weforum.org/agenda/2020/12/anti-corruption-transparency-restore-trust-in-leadership/

Bruckner. (2019). *2020–30: The Decade of anti-corruption?* https://www.thelancet.com/pdfs/journals/langlo/PIIS2214-109X(19)30500-5.pdf.

Campbell, B. (2012). Corporate social responsibility and development in Africa: Redefining the roles and responsibilities of public and private actors in the mining sector. *Resources Policy, 37*(2), 138–143. https://www.sciencedirect.com/science/article/pii/S0301420711000377

Cohen, S. (2001). A strategic framework for devolving responsibility and functions from government to the private sector. *Public Administration Review, 61*(4), 432–440.

Corcuera-Santamaria, S., & Leonor Moreno, M. (2019). *How blockchain can make trade safer.* IDB. https://blogs.iadb.org/integration-trade/en/blockchain-trade-safer/

Cosgrove, E. (2018). *9 ocean carriers, terminal operators join new blockchain initiative to rival TradeLens.* SupplyChainDive. https://www.supplychaindive.com/news/ocean-carriers-new-blockchain-cosco-cma-cgm/541630/

Davidson Raycraft, R., & Lannquist, A. (2020). How governments can leverage policy and blockchain technology to stunt public corruption. *World Economic Forum.* https://www.weforum.org/agenda/2020/06/governments-leverage-blockchain-public-procurement-corruption/

De Meijer, R. W. C. (2018). *African countries open for blockchain acceptance.* Finextra. https://www.finextra.com/blogposting/15656/african-countries-open-for-blockchain-acceptance.

dhs.gov. (2017). *News release: DHS S&T Awards $750K to Virginia Tech company for blockchain identity management research and development.* Homeland Security. https://www.dhs.gov/science-and-technology/news/2017/09/25/news-release-dhs-st-awards-750k-virginia-tech-company

Duque, M. S. (2017). *En un año se robaron 32,8 millones de raciones de comida del pae.* El Tiempo. https://www.eltiempo.com/justicia/investigacion/desfalco-de-32-8-millones-en-raciones-de-comida-escolar-de-los-ninos-153928.

Dwyer, G. (2016). Blockchain: A Primer, MPRA Paper 76562, University Library of Munich.

Edward-Ekpu, U. (2020). *Nigeria is now the No.2 bitcoin market on this fast-growing global marketplace.* Quartz Africa. https://qz.com/africa/1947769/nigeria-is-the-second-largest-bitcoin-market-after-the-us/

eltiempo.com. (2017). Una pechuga de pollo a $ 40.000 y huevo a $900, en sobrecostos del PAE. El Tiempo. https://www.eltiempo.com/justicia/investigacion/sobrecostos-en-programa-de-alimentacion-escolar-en-colombia-153590

ENS. (2019). Andhra government to adopt blockchain tech to end land record tampering. *The New Indian Express.* https://www.newindianexpress.com/states/andhra-pradesh/2019/dec/15/andhra-government-to-adopt-blockc hain-tech-to-end-land-record-tampering-2076359.html

Finer, H. (1941). Administrative responsibility in democratic government. *Public Administration Review, 1*(4), 335–350.

Fox, J. (1994). Latin America's emerging local politics. *Journal of Democracy, 5*(2), 105–116.

Gabison, G. (2016). Policy considerations for the blockchain technology public and private applications. *SMU Science and Technology Law Review, 19*, 327–350. https://heinonline.org/HOL/LandingPage?handle=hein.journals/com lrtj19&div=19&id=&page=

Giraldo, S. C. (2019). *Peru's government looks to blockchain to fight corruption.* Perú Reports. https://perureports.com/perus-government-looks-to-blo ckchain-to-fight-corruption/9045/.

Global Banking & Finance Review. (2018). Bank and blockchain in Africa: A lot of African banks and government run when they hear the word crypto because of the effect they feel it will have on the economy and also loss of control. https://www.globalbankingandfinance.com/bank-and-blockc hain-in-africa-a-lot-of-african-banks-and-government-run-when-they-hear-the-word-crypto-because-of-the-effect-they-feel-it-will-have-on-the-economy-and-also-loss-of-control/

Goel, R., & Nelson, M. (2007). Are corrupt acts contagious? Evidence from the United States. *Journal of Policy Modeling, 29*(6), 839–850. https://www.sci encedirect.com/science/article/pii/S0161893807001056

Gray, C. W., & Kaufman, D. (1998). Corruption and development. PREM Notes; No. 4. World Bank©. https://openknowledge.worldbank.org/han dle/10986/11545 License: CC BY 3.0 IGO.

Greenfield, R. (2017). *Using blockchain to prevent police corruption & brutality.* Medium. https://robertgreenfieldiv.medium.com/using-blockchain-to-pre vent-police-corruption-brutality-c998d5ea0732

Guzzo. (2020). *Observations regarding the use of blockchain by the Brazilian Public Administration*. Souto Correa Advogados. https://www.soutocorrea. com.br/en/artigos/observations-regarding-the-use-of-blockchain-by-the-bra zilian-public-administration/.

Guest Contributors More articles by this source. (2018). *An overview of Latin America's blockchain adoption*. Nasdaq. https://www.nasdaq.com/articles/ overview-latin-americas-blockchain-adoption-2018-06-20.

Hall, I. (2020). *Colombian blockchain trial cause for 'cautious optimism', says WEF*. Global Government Forum. https://www.globalgovernmentforum. com/colombian-blockchain-trial-cause-for-cautious-optimism-says-wef/.

Hartson, W. (2016). *Top 10 facts about corruption*. Express.co.uk. https:// www.express.co.uk/life-style/top10facts/669514/top-ten-facts-corruption-Tackling-Corruption-Together-conference-London.

Hartzenberg, T. (2011). Regional Integration in Africa, SSRN. https://papers. ssrn.com/sol3/papers.cfm?abstract_id=1941742.

Heeks, R. (1998, February 18). Information technology and public sector corruption. *Information Systems for Public Sector Management Working Paper no. 4*, SSRN. https://ssrn.com/abstract=3540078 or https://doi.org/10. 2139/ssrn.3540078

Higgins, S. (2016a). *Republic of Georgia to develop blockchain land registry*. Coindesk. http://www.coindesk.com/bitfuryworking-with-georgian-government-on-blockchain-land-registry/

Higgins, S. (2016b). *Survey: Blockchain capital markets spending to reach $1 billion in 2016*. Coindesk. http://www.coindesk.com/capital-markets-1-bil lion-2016-blockchain/

Hope, K., & Chikulo, B. (2000). *Corruption and Development in Africa*. Palgrave Macmillan. https://www.palgrave.com/gp/book/9780333770894

Hudon, P., & Garzón, C. (2016). Corruption in public procurement: Entrepreneurial coalition building. *Crime, Law and Social Change, 66*(3), 291–311, Springer Link. https://link.springer.com/article/10.1007/s10 611-016-9628-4

Insights, L. (2020a). Indonesia customs joins TradeLens shipping blockchain platform, Ledger Insights—Blockchain for enterprise. https://www.ledgerins ights.com/tradelens-enterprise-blockchain-shipping-indonesia-customs/.

Insights, L. (2020b). *Standard chartered joins TradeLens shipping blockchain platform*. Ledger Insights—Blockchain for enterprise. https://www.ledgerins ights.com/standard-chartered-tradelens-shipping-blockchain/.

Insights, L. (2020c). *UN world food programme uses blockchain for direct payments*. Ledger Insights - blockchain for enterprise. https://www.ledgerins ights.com/un-world-food-programme-uses-blockchain-for-direct-payments/.

Isaac, K. (2019). Kenya loses Ksh.1 trillion every year to corruption—what could 6 trillion stolen so far in jubilee government do? Soko Directory. https://sok odirectory.com/2019/02/kenya-loses-1-trillion-every-year-corruption/

Jagati, S. (2020). *Blockchain interoperability: The holy grail for Cross-Chain deployment*. Cointelegraph. https://cointelegraph.com/news/blockchain-int eroperability-the-holy-grail-for-cross-chain-deployment.

Johnson. (2018). Vegetarianism is good for the economy too. *World Economic Forum*. https://www.weforum.org/agenda/2018/12/vegetaria nism-is-good-for-the-economy-too/.

Jones, G. W. (1992). The search for local accountability. In Leach, S. (Ed.), *Strengthening local government in the 1990's* (pp. 49–78).

Jones, A. (2016). *How blockchain is impacting industry*. International Banker. http://internationalbanker.com/finance/blockchain-impacting-industry/

Kaaru, S. (2020). *Un calls for blockchain to fight rampant corruption in Kenya*. CoinGeek. https://coingeek.com/un-calls-for-blockchain-to-fight-rampant-corruption-in-kenya/.

Kapadia, S. (2018). *Blockchain solution promises to save millions for ocean freight*. Supply Chain Dive. https://www.supplychaindive.com/news/blockc hain-ocean-freight-savings/519237/.

Klitgaard, R. (1988). *Controlling Corruption*. University of California Press.

Kohli, J. (2012). Red flags in E-Procurement / E-Tendering for public procurement and some remedial measures. Kohli, J. (2010), E-Procurement integrity matrix, transparency international India; OECD. (2018). Mexico's E-Procurement system: Redesigning compranet through stakeholder engagement, OECD Publishing.

Kriticos. (2019). *Keeping it clean: Can blockchain change the nature of land registry in developing countries?* World Bank Blogs. https://blogs.worldbank. org/developmenttalk/keeping-it-clean-can-blockchain-change-nature-land-registry-developing-countries.

Kshetri, N. (2017a). Blockchain's roles in strengthening cybersecurity and protecting privacy. *Telecommunications Policy, 41*(10), 1027–1038.

Kshetri, N. (2017b). Will blockchain emerge as a tool to break the poverty chain in the Global South? *Third World Quarterly, 38*(8), 1710–1732.

Kshetri, N. (2018). Blockchain could be the answer to cybersecurity. Maybe. *Wall Street Journal*. https://www.wsj.com/articles/blockchain-could-be-the-answer-to-cybersecurity-maybe-1527645960

Kshetri, N., & Voas, J. (2018). Blockchain in developing countries. *IEEE IT Professional, 20*(2), 11–14.

Landau, M. (1991). On multiorganizational systems in public administration. *Journal of Public Administration Research and Theory: J-PART, 1*(1), 5–18.

Las aduanas del mercosur están conectadas por blockchain. (2020). Latam. https://latam.portalerp.com/las-aduanas-del-mercosur-estan-conectadas-por-blockchain.

Lemieux, V. (2016). Trusting records: Is blockchain technology the answer? *Records Management Journal, 26*(2), 110–139.

Lemieux, V. (2017). Evaluating the use of blockchain in land transactions: An archival science perspective. *European Property Law Journal.* https://www.degruyter.com/view/j/eplj.2017.6.issue-3/eplj-2017-0019/eplj-2017-0019.xml

LoadstarEditorial. (2020). *South Asia gateway terminals becomes first Sri Lankan terminal to join TradeLens to digitalise supply chains.* The Loadstar. https://theloadstar.com/south-asia-gateway-terminals-becomes-first-sri-lankan-terminal-to-join-tradelens-to-digitalise-supply-chains/.

López, M. (2019). *Brazil wants to build customs connectivity with Blockchain.* Contxto. https://contxto.com/en/brazil/brazil-build-customs-connectivity-blockchain/.

Mary, L., & Remko, V. H. (2021). *What we've learned so far about blockchain for business.* MIT Sloan Management Review. https://sloanreview.mit.edu/article/what-weve-learned-so-far-about-blockchain-for-business/.

Maltaverne. (2021). *What can blockchain do for public procurement?* Public Spend Forum. https://www.publicspendforum.net/blogs/bertrand-maltaverne/2017/08/28/blockchain-technology-public-procurement/.

Manski, S. (2017). Building the blockchain world: Technological commonwealth or just more of the same? *Strategic Change, 26*(5), 511–522.

Mawenya, A. S. (2008). Preventing corruption in African procurement, Africa Portal. South African Institute of International Affairs (SAIIA). https://www.africaportal.org/publications/preventing-corruption-in-african-procurement/.

Mazor, Y. (2018). *China fighting tax corruption with blockchain.* FXEmpire. https://www.fxempire.com/news/article/china-fighting-tax-corruption-with-blockchain-522578

Mbaku, J. M. (1996). Bureaucratic corruption in Africa: The futility of cleanups. *Cato Journal, 16*(1), 99–118.

Mendes, K. (2018). *Can blockchain save the Amazon in corruption-mired Brazil?* Reuters. Thomson Reuters. https://www.reuters.com/article/us-brazil-property-blockchain/can-blockchain-save-the-amazon-in-corruption-mired-brazil-idUSKBN1FE113.

MIT Technology Review. (2017). *Understand why Ethereum exists, and you'll get why it's a big deal.* https://medium.com/mit-technology-review/understand-why-ethereum-exists-and-youll-get-why-it-s-a-big-deal-df6765a5805d

Morris, N. (2019). Hapag-Lloyd, one join IBM Maersk TradeLens shipping blockchain, Ledger Insights - blockchain for enterprise. https://www.ledger insights.com/hapag-lloyd-one-ibm-maersk-tradelens-shipping-blockchain/.

Mota, R. (2020). Brazil regulates blockchain in foreign trade systems. Olhar Digital. https://olhardigital.com.br/en/2020/11/30/pro/brasil-regula menta-blockchain-nos-sistemas-de-comercio-exterior/?gfetch=2020%2F11% 2F30%2Fpro%2FBrazil-regulates-blockchain-in-foreign-trade-systems%2F

Mulgan, R. (2000). Accountability: An ever-expanding concept. *Public Administration, 78*(3), 555–573.

NEC, IDB Lab and NGO Bitcoin Argentina to deploy a blockchain-ba. (2019). Bloomberg.com. Bloomberg. https://www.bloomberg.com/press-releases/ 2019-08-26/nec-idb-lab-and-ngo-bitcoin-argentina-to-deploy-a-blockchai n-ba.

Newham, & Faull. (2011). Reflections of academic and professional leaders on leadership in a …, Sabinet. https://journals.co.za/doi/10.10520/ejc-sajhrm-v18-n1-a29.

OECD. (2014). OECD Foreign Bribery Report. OECD. https://www.oecd-ili brary.org/governance/oecd-foreign-bribery-report_9789264226616-en;jsessi onid=kqkNGs7XJt27K6iV6H4h5xNA.ip-10-240-5-102

OECD. (2016). Preventing corruption in public procurement. OECD. http:// www.oecd.org/gov/ethics/Corruption-Public-Procurement-Brochure.pdf

Ogundeji, O. (2016). *Land registry based on blockchain for Africa.* http:// www.itwebafrica.com/enterprise-solutions/505-africa/236272-land-registry-based-on-blockchain-for-africa.

Osterfeld, D. (1992). *Prosperity versus planning. How government stifles economic growth.* Oxford University Press.

Paynter, B. (2017). *How blockchain could transform the way international aid is distributed.* Fast Company. www.fastcompany.com/40457354/how-blockc hain-could-transform-the-way-international-aid-is-distributed.

Piga, C. (2011). *A fighting chance against corruption in public procurement?* EconPapers. https://econpapers.repec.org/bookchap/elgeechap/14003_ 5f5.htm

Promise and peril: Blockchain, Bitcoin and the fight against… .(no date). Transparency.org. https://www.transparency.org/en/news/promise-and-peril-blo ckchain-bitcoin-and-the-fight-against-corruption.

Ramnani, V. (no date). *Transfer of property title likely on same day through blockchain.* Moneycontrol. https://www.moneycontrol.com/news/business/ real-estate/transfer-of-property-title-likely-on-same-day-through-blockchain-2449981.html.

Raymond, J. (2008). Benchmarking in public procurement, *Benchmarking: An International Journal, 15*(6), 782–793. https://doi.org/10.1108/146357 70810915940

REUTERS. (2021). Bitcoin donations surge to jailed Kremlin critic Navalny's cause, The Jerusalem Post | JPost.com. https://www.jpost.com/breaking-news/bitcoin-donations-surge-to-jailed-kremlin-critic-navalnys-cause-658766.

Romzek, B. (2012). A preliminary theory of informal accountability among network. *Public Administration Review.* https://onlinelibrary.wiley.com/doi/full/10.1111/j.1540-6210.2011.02547.x.

Royal Malaysian Customs Department adopts IBM and Maersk's TradeLens Blockchain. (no date). *TradeLens.* https://www.tradelens.com/press-rel eases/royal-malaysian-customs-department-adopts-ibm-and-maersks-tradel ens-blockchain.

Rustiarini, N. W., Sutrison, T., Nurkholis, N., & Andayani, W. (2019). Why people commit public procurement fraud? The fraud diamond view. *Journal of Public Procurement, 19*(4), 345–362. https://doi.org/10.1108/JOPP-02-2019-0012

Stanley, A. (2021). *Microlending startups look to blockchain for loans, CoinDesk Latest Headlines RSS.* CoinDesk. https://www.coindesk.com/microlending-trends-startups-look-blockchain-loans.

Swan, M. (2017). Anticipating the economic benefits of blockchain. *Technology Innovation Management Review, 7*(10).

Tapscott, D., & Ticoll, D. (2003). *The naked corporation.* The Free Press.

Thontteh, E. O., & Omirin, M. M. (2015). Land registration within the framework of land administration reform in Lagos state. *Pacific Rim Property Research Journal, 21*(2), 161–177.

Till, B.M. et al. (2017). Blockchain and global health, *Foreign Affairs.* https://www.foreignaffairs.com/articles/world/2017-11-03/blockchain-and-global-health.

Tschantz, M. S., & Ernst, M. D. (2005). Javari: Adding reference immutability to Java. In *Proceedings of the 20th Annual ACM SIGPLAN Conference on Object-Oriented Programming, Systems, Languages, and Applications* (pp. 211–230).

UN News Center. (2011). Corruption leading to unequal access, use and distribution of land—UN report. UN. http://www.un.org/apps/news/story.asp?NewsID=40698#.WEMpP33QCWl.

UNCTAD. (2020). "Policy brief." UNCTAD. https://unctad.org/system/files/official-document/presspb2020d1_en.pdf

Verkijika, S., & De Wet, L. (2018). A usability assessment of e-government websites in Sub-Saharan Africa. *International Journal of Information Management, 39,* 20–29. https://www.sciencedirect.com/science/article/abs/pii/S0268401217307934

Vos, J., Lemmen, C., & Beentjes, B. (2017). Blockchain-based land administration: Feasible, illusory or panacae?" In *Paper Prepared for Presentation at the Annual World Bank Conference on Land and Poverty, 2017* (The World Bank).

weforum.org. (2020). Exploring blockchain technology for government transparency: Blockchain-based public procurement to reduce corruption. *World Economic Forum*. http://www3.weforum.org/docs/WEF_Blockchain_Gove rnment_Transparency_Report.pdf

Wong, J. I. (2017). *The UN is using Ethereum's technology to fund food for thousands of refugees*. Quartz. https://qz.com/1118743/world-food-progra mmes-ethereum-based-blockchain-for-syrian-refugees-in-jordan/.

Wood, M. (2020). Thai customs joins TradeLens blockchain platform. *Ledger Insights*. https://www.ledgerinsights.com/thai-customs-joins-tradel ens-blockchain-platform/

xinhuanet.com. (2019). *China reforms judicial courts using internet technologies: White paper*. XinhuaNet. http://www.xinhuanet.com/english/2019-12/05/ c_138605955.htm

Young, T. (2021). *Blockchain technology cuts through the hurdles to simplify everyone's lives*. The National. https://www.thenationalnews.com/business/ blockchain-technology-cuts-through-the-hurdles-to-simplify-everyone-s-lives-1.85148.

CHAPTER 3

Facilitation of Entrepreneurial Activities

Abstract This chapter examines the potential roles of blockchain in facilitating entrepreneurial activities in the Global South by dismantling the current core–periphery model. It provides a detailed analysis and description of how blockchain-based financial technologies and cryptocurrencies can make it possible for entrepreneurs to connect with investors, which can facilitate small entrepreneurs' access to low-cost loans. It also demonstrates how blockchain facilitates small firms' access to supply chain and trade finance by providing lenders with high-quality and relevant data about potential borrowers and their activities. It discusses how blockchain can help small entrepreneurial firms in the Global South to demonstrate product quality. The chapter also addresses how entrepreneurial firms can combine blockchain with other technologies to demonstrate that they protect workers from poor labor conditions such as child labor, forced labor, and bad working conditions.

Keywords Creditworthiness · Cryptocurrencies · Economies of scale · Entrepreneurial activities · Small and medium-sized enterprises · TechFins

3.1   INTRODUCTION

Individual entrepreneurs and entrepreneurial firms in developing economies face significant barriers in raising finance to start their businesses and accessing other factors of production and resources such as markets for their products. Especially unavailability of financing is among the most critical barriers faced by most entrepreneurs in developing countries (Kshetri, 2017). For instance, financial systems in Africa are "small, shallow and costly" and enterprises in the region have rated access to finance as the most significant constraint on their operation and growth compared to those outside Africa (Beck & Cull, 2014).

When it comes to entrepreneurship-related activities, firms in developing countries often experience peripherality. Note that peripherality is defined as the condition experienced by individuals, firms, and economies that keeps them away from the global economy's controlling center (Goodall, 1987). For instance, venture capital investments in 2017 in Africa were US$0.56 billion in Africa compared to US$74.5 billion in North America (Kshetri, 2019). This is because most venture capitalists are located in North America (the controlling center) and developing economies such as those in Africa are located in the periphery.

Likewise, there is a lack of a level playing field for developing world-based small and medium-sized enterprises (SMEs) in terms of market access (Van der Meer, 2006), *both*domestic and international (Kshetri, 2021). For many developing world-based manufacturers, compliance with Environmental, Social, and Governance (ESG) standards is a prerequisite for access to the developed world market. In most cases, the burden of proof for determining that no ESG *violation has* occurred in the production process is high and often lies with the developing world-based firms. For instance, *artisanal* and small-scale miners (ASMs) in Rwanda are required to pay between US$780 to US$1080 to use *ITRI* Tin Supply Chain Initiative (ITSCI) traceability system. These charges are prohibitively expensive for subsistence miners such as ASMs (Kshetri, 2021).

Blockchain-based solutions are also likely to result in a reduction in costs to access international markets for developing world-based SMEs. The blockchain-based traceability-as-a-Service (TaaS) provided Circulor has said that its system will change the business models of ASMs by shifting traceability costs from miners to end users (Mwai, 2018).

Due to falsifications *such as fake green certifications and labels* and counterfeit certifications of organic processes or farm inspections by some fraudsters (Sustainable Food News, 2018), even legitimate producers of genuine products from developing countries face difficulties in entering the global market. Blockchain, in combination with other technologies, has a potential to address such challenges. Some companies are also developing solutions involving blockchain, satellite images and artificial intelligence (AI) that reward sustainable farming practices. One such example is database management company Oracle's partnership with the World Bee Project to help farmers manage bee populations and pollinator habitats. The plan is to take images of the farm with drones or satellites and utilize AI-based image recognition algorithms to evaluate whether the way a farmland is managed support bee colonies and other pollinators in a sustainable way (Charness, 2019). Research has indicated that farms that allocate a certain proportion of their land to plant flowering crops such as spices, oil seeds, buckwheat, and sunflowers can increase crop yields by up to 79% due to efficient pollination from bees. An eco-label certificate can be issued to farmers depending on the farm composition. The certification can be stored in a blockchain so that all supply chain partners can see it (e.g., during the farm product's journey to the retailer).

In terms of entrepreneurial activity and performance, being farther from the center of the action, GS-based entrepreneurial firms face challenges in accessing financing sources such as venture capital investors and finding new markets for their products. *Separated by space and time,* the periphery functions as a vassal of the core and is associated with marginal activities such as low-value primary production (Anderson, 2000). Blockchain-based solutions have the potential to change the peripheral status of developing world-based entrepreneurial firms.

3.2 Key Issues in Entrepreneurial Settings in the GS

There are three sections in this literature review. The first two sections deal with challenges associated with the access to market, finance, and other resources. In the last section, we look at ICTs' potential roles in improving entrepreneurship.

Access to and Costs of Finance

Access to finance has been a major problem facing enterprises in Africa (Fombang & Adjasi, 2018) and developing countries in other regions (Kshetri, 2019). Among the major challenges that potential entrepreneurs face in developing economies is the lack of clear property rights. For instance, about 90% of the land is undocumented or unregistered in rural Africa. Likewise, the lack of land ownership remains among the most important barriers to entrepreneurship in India (Kshetri, 2016). One estimate suggested that over 20 million rural families in India did not own land and millions more lacked legal ownership to the lands they built houses on, lived on, and worked (Hanstad, 2013). Indeed, landlessness is arguably a more powerful predictor of poverty in India than caste or illiteracy. A large proportion of poor people in the developing world lack property rights. This means that many potential entrepreneurs cannot use their assets as collateral to increase access to capital.

Regarding business *loans without collateral*, prior researchers have identified two main problems in emerging economies: inefficiency and informational opacity. First, traditional banks are unwilling and reluctant to serve small-scale borrowers such as poor people and small businesses due to high transaction costs and inefficient processes associated with making small loans to these borrowers (Adams & Nehman 1979). The second reason why poor people and small businesses face barriers to accessing financial products concerns *informational opacity* (Stiglitz & Weiss, 1981). A national credit bureau would collect and distribute reliable credit information, and hence increase transparency and minimize the banks' lending risks. However, many Southern economies are characterized by the *lack, or poor performance, of credit rating agencies* to *provide* information about the *creditworthiness* of SMEs. This situation puts SMEs in a disadvantaged position in the credit market. SMEs tend to be more informationally opaque than large corporations because the former often lack certified audited financial statements and thus it is difficult for banks to assess or monitor the financial conditions (Kshetri, 2019).

After-sales payment is another challenge faced by small-scale producers. For instance, small farmers that supplied to *Johannesburg Fresh Produce Market* claimed that marketing agencies did not pay them on time after their produce was sold. This negatively affected their cash flow and led to further problems in production and market access (Bbun & Thornton, 2013).

Access to Market and Other Resources

Developing world-based small firms have limited access to factors of production such as working capital, credit, and information; and markets (Ortmann & King, 2007). These firms confront various barriers and obstacles to accessing markets for their products. Van der Meer (2006) has noted that the lack of economies of scale hinders small-scale farmers' capability to compete against large-scale farmers in the food markets worldwide (Van der Meer, 2006). In Bbun and Thornton's (2013) study of urban and peri-urban agriculture (UPA) in South Africa, 36% of small-scale producers reported that they were more disadvantaged in market access compared to large-scale farmers.

The limited availability of *data about quality-related* outcomes hinders small farmers' entrepreneurial success. This is mainly because the official grading of farm produce often takes place only after the produce reaches the formal market. However, grains are likely to suffer quality loss before they reach the formal market, and such loss often goes unrecorded. Informal markets often exhibit a lower *level of quality consciousness* than their formal counterparts. Furthermore, formal markets across different countries operate different grading systems for each grain type. The measures of grain quality are not thus equivalent, which means that quality-related indicators cannot be translated from one market to the other (Hodges, 2012).

ICTs and Entrepreneurship

Two key attributes—agency and boundaries—provide a helpful theoretical perspective for understanding ICTs' roles in changing the organizational and inter-organizational processes and outcomes (Nambisan, 2017). Agency refers to the "capacity for action" (Giddens, 1984). In an agency relationship, an "agent" is designated to represent the "principal". The agent takes action or makes decisions on behalf of the principal (Ross, 1973). It is argued that ICTs function as a material agency since they can be assigned to perform actions without direct or complete control of human beings (Faulkner & Runde, 2009; Orlikowski & Scott, 2008). To take an example, digital platforms such as crowdsourcing and crowdfunding systems enable a group of actors to jointly create value (Nambisan, 2017).

When ICTs function as a material agency, they may change the boundary of actions that human agents can take. ICTs have also made entrepreneurial processes and outcomes less bounded. For instance, entrepreneurial activities are less restricted in terms of temporal and spatial dimensions.

von Briel et al. (2018) examined ICTs' two key characteristics—specificity and relationality—that affect agency and boundaries. A high degree of specificity and control over actions would increase the predictability of inputs, which can reduce the variance in outputs produced.

The specificity feature of ICTs can be described in terms of restrictiveness (set of actions that can be possibly performed), comprehensiveness (number of features offered by a technology) DeSanctis and Poole (1994) and adaptivity (set of actions and interactions enabled by the technology) (von Briel et al., 2018).

The relationality property involves ICTs' relationships with other actors that enhance their functionality (Kallinikos et al., 2013). Relationality influences the nature and number of actors that can participate in the venture creation processes (von Briel et al., 2018).

ICTs vary widely in terms of the capacity for relationality. For example, a 3D printer is characterized by a low degree of relationality as it mostly connects with a single actor at a time. On the other hand, social media have a high degree of relationality since they can connect diverse participants (von Briel et al., 2018).

In addition to the number of actors, it is important to look at the nature and quality of relations that exist between them. While some ICT applications such as social media are characterized by a high degree of relationality (von Briel et al., 2018), such relationships are often shallow. A key feature of shallow relationships such as those observed in most social media applications is that there is a high level of uncertainty. Without the perceptions of integrity, the trust required for coordination and cooperation may not exist (Pirson & Malhotra, 2011). One way would be to rely on transactional *trust* (Zucker, 1988).

By connecting more diverse actors with complementary capabilities, ICTs increase their potential to create new combinations of resources. These actors can also engage in the modification of resources to engage in value-creation activities (von Briel et al., 2018).

Blockchain and Entrepreneurship

Prior researchers have identified various mechanisms by which blockchain can facilitate entrepreneurship. Block et al. (2018) noted that blockchain and cryptocurrencies have the potential to change the business models of existing players in the entrepreneurial finance area. Among the major benefits, these technologies provide novel ways to assess risks and evaluate financial information. Cryptocurrencies can also make it possible for non-professional investors to participate in entrepreneurial financing. Additional benefits include greater liquidity, and reduction in monitoring costs for investors (Block et al., 2018).

A rapidly evolving phenomenon of particular interest is a capital-raising method known as an initial coin offering (ICO). ICO involves offers and sales of cryptotokens using blockchain (Kshetri, 2023).

3.3   Blockchain's Roles in Facilitating Entrepreneurial Activities in the Global South

In Table 3.1, we present various mechanisms by which blockchain can *facilitate entrepreneurial activities*. The table points out two broad categories of applications of blockchain: n*on-cryptocurrency applications* and cryptocurrency *applications.*

Table 3.1   The uses of blockchain in facilitating entrepreneurial activities

Type of application → Phases of entrepreneurial activity	Non-cryptocurrency applications	Cryptocurrency applications
Starting a business	• Authentic information to use assets as a collateral to get a loan. • Relevant and authentic information to access potential borrowers' creditworthiness	• Efficient and cost-effective way to raise capital
Raising and operating a business to maximize profits	• Facilitating the access to factors of production such as working capital • Demonstrating firms' and products' compliance with various regulations, standards and other requirements	• Overcoming geopolitical disadvantage in accessing key resources and handling business transactions

Cryptocurrencies are digital or virtual currencies built with strong cryptography to make them secure and immutable. Most cryptocurrencies are built using blockchain (O'Neal, 2019). Blockchain-based cryptocurrencies can be broadly divided into two types: a coin and a token. *Digital coins or* cryptocurrency coins *are native to* or *created using their own blockchain. Some examples include* Bitcoin, Monero, and Ether. Tokens, on the other hand, require another blockchain platform to create and operate. A cryptotoken is a unit of value issued by a project or company, which rewards token owners. It allows the owner to perform particular actions (e.g., get a specific service on the network).

We look at two phases of entrepreneurial activities: a) Starting a business and b) Raising and operating a business to maximize profits. The subheadings in this section are organized according to Table 3.1. In Table 3.2, we look at how blockchain performs in terms of two key features– specificity and relationality—in terms of its roles in facilitating entrepreneurial activities.

Table 3.2 Specificity and relationality of blockchain and facilitation of entrepreneurial activities

Characteristic Phases of entrepreneurial activity	Specificity	Relationality
Starting a business	• Ensuring that the asset used as a collateral is not fraudulent. • Relevant and authentic information to assess potential borrowers' creditworthiness: increases the predictability of the loan repayment by a borrower	• Facilitates P2P lending by directly connecting lenders and borrowers thereby eliminating the need for intermediaries.
Raising and operating a business to maximize profits	• Firms' and products' compliance with various regulations, standards and other requirements can be verified with a high degree of certainty (high specificity) • Eliminates the possibility of fraudulent invoices (high specificity)	• Cryptocurrencies' decentralization can dismantle the core-periphery relation: Entrepreneurs in the Global South do not suffer from the peripherality status

Authentic Information to Use Assets as Collateral to Get a Loan

Among the key reasons for the lack of access to finance in developing countries (Fombang & Adjasi, 2018), many financial institutions in these countries have become victims of fraudulent acts and so are reluctant to give loans to entrepreneurs. Fraudulent activities are especially likely to be committed by individuals with political connections. Such individuals are less likely to be the *target of investigation and enforcement* (Cumming et al., 2006). The failure of accounting firms to detect managerial fraud has also led to less faith in audited financial statements. Worse still, many believe that the accounting firms have compromised their own integrity because of the lure of lucrative consulting contracts from firms they were auditing (Zahra et al., 2007).

By using blockchain solutions, fraudulent activities such as those found in *Qingdao* can be reduced. Indeed, recent high-profile frauds have increased blockchain's attractiveness. The *British* multinational banking and financial services company, Standard Chartered lost about US$200 million from Qingdao frauds. Standard Chartered teamed up with DBS Group and Singapore's Infocomm Development Authority to develop a blockchain-based platform (Chanjaroen & Boey, 2016). Standard Chartered is a participating bank in blockchain-based trade finance platforms such as eTradeConnect and Bay Area Trade Finance Blockchain Platform.

To take another example, India's central bank, the Reserve Bank of India licensed three entities RXIL, A.TReDS, and M1xhange to provide receivable financing to micro and small businesses. These three platforms wanted to share information in order to prevent fraud but keep the data private. Blockchain can help achieve this. Using blockchain it is possible to create a cryptographic representation of the invoice, known as a hash. A hash provides an indecipherable text and does not tell anything about the invoice. It is nearly impossible to convert a hash back to the original data. If a trader submits the same invoice to more than one trade finance platform, the hash will match, which raises a red flag. New York-based MonetaGo also hashes some of the elements of the invoice in order to prevent the trader from making some modifications to the invoice. An invoice that has a high degree of similarity with another invoice already submitted to a different platform will not be rejected, but it will produce an amber flag. The platform may then ask the trader to explain (Morris, 2018). As of September 2020, MonetaGo made over one million transactions in India.

Blockchain solutions launched by MonetaGo eTradeConnect and Bay Area Trade Finance Blockchain Platform ensure that the *asset* used as collateral *is not fraudulent*. In this way, blockchain can be considered to be a technology with high specificity (von Briel et al., 2018).

Relevant and Authentic Information to Access Potential Borrowers' Creditworthiness

Developing economies lack information about the *creditworthiness* of most of the population.

Large non-financial firms, especially from technology sectors—also referred to as TechFins—are leveraging their data asset aggressively in order to provide financial services (Zetsche et al., 2018). In many cases, however, TechFins' datasets originate from sources such as social media that are unrelated to financial services. Big data analytics used to predict potential borrowers' creditworthiness rely on correlations rather than causations. It is not easy to determine which correlations shown by big data tools are random and which ones may reflect responsible financial and consumption decisions (Zetsche et al., 2018).

Even more importantly, financial and other forms of exclusions are more likely to affect non-users of certain services such as social media (Zetsche et al., 2018). Compared to wealthy individuals, larger proportions of poor people are non-users of cellphones, the Internet, and social media, which are used by TechFins and FinTech companies to analyze creditworthiness. For instance, in the least developed countries (LDCs), which are low-income countries that perform poorly in human assets and face high economic vulnerability, more than 27% of the population did not have cellphones and more than 80% lacked Internet access in 2019.

A key benefit of blockchain-based applications concerns the availability and utilization of more relevant information to assess creditworthiness. For instance, many women, small-scale farmers, migrant workers, refugees and displaced people receive foreign aid and humanitarian assistance from the United Nations or international non-governmental organizations. Some of them receive microfinance loans. Others participate in various training programs. The different categories of information are often stored in independent disjoint databases such as those of microfinancing institutions, the UN and relevant INGOs (Ramirez, 2017). Such data are unlikely to be captured by TechFins' algorithms.

Some blockchain systems aim to capture and organize more relevant information compared to those used by TechFins. For instance, the microlending platform Kiva's blockchain system captures all credit-related events in a single ledger connected to a borrower's wallet. Likewise, the U.S.-based blockchain company BanQu's "economic passport" aggregates financial history and other information from a number of sources (Stanley, 2017). These blockchain systems use more fundamental rather than proxy attributes to characterize a potential borrower's ability and willingness to pay. Likewise, future plans for the World Food Program's (WFP) blockchain program Building Blocks is to allow refugees and displaced people to control their own cryptographic keys and integrate personal data sources such as the WHO, the United Nations Children's Fund (UNICEF), and the WFP in order to build their economic identity and creditworthiness.

BanQu utilizes Ethereum blockchain to establish economic identities and proofs of record (economic passport) (Stanley, 2017). It aggregates information from a number of sources such as those related to financial history, land records, trust networks documenting trust relationships with others and business registrations, vaccination records, and remittance income. ID-related information sources include selfies, biometrics, and key physical attributes. It also provides Know Your Customer (KYC) and other information to the partners that can potentially offer products and services to these disadvantaged individuals.

BanQu has teamed up with the multinational drink and brewing holdings company Anheuser-Busch InBev to promote supply chain transparency and traceability in India, Uganda, and Zambia. The systems track cassava and barley supplied to Anheuser-Busch. The BanQu system is also referred to as Chembe cassava online buying project in Zambia. The partnership started with the cassava crop value chain with an aim to provide economic empowerment to small-scale farmers. Using BanQu solutions, Anheuser-Busch InBev's local business, Zambian Breweries can track its products throughout the supply chain: from the farmer to local businesses to aggregated buyers and retailers. BanQu uses GPS to locate farmers. The located farmers are identified by agents to facilitate and verify transactions. Geo-location tags and farmers' identity profiles and other pieces of information are put on the blockchain (https://www.craftbrewingbusiness.com/news/blockchain-breakthrough-poor-zambian-farmers-are-now-empowered-within-ab-inbevs-supply-chain/). The system has all the information to know who supplied what and when.

The discussion in this section makes it clear that relevant and authentic information to access potential borrowers' creditworthiness would *increase the predictability of the loan* repayment by a borrower. Blockchain-based information thus performs well in terms of specificity property (von Briel et al., 2018).

Facilitating Access to Factors of Production Such as Working Capital

As noted above, small firms in developing countries have limited access to factors of production such as working capital and credit (Ortmann & King, 2007). They experience difficulties in meeting working capital requirements due to the delay in the payment of trade receivables. Especially small businesses are often short of cash and find it expensive—if not impossible—to borrow money.

Several blockchain-based platforms have been developed to fulfill entrepreneurial firms' trade finance and supply chain finance needs in developing countries. As an example, in 2018, China's central bank, the People's Bank of China (PBoC) started to pilot a trade finance (TF) project referred to as the Bay Area Trade Finance Blockchain Platform (BATFB). The solution targets SMEs. The BATFB was launched in Shenzhen. Targeted jurisdictions included Guangdong, Hong Kong, and Macau. The banks participating in the project include Bank of China, China Construction Bank, China Merchants Bank, Ping An Bank, and Standard Chartered Bank (Faridi, 2020).

In May 2020, the BATFB received $4.7 million in "special funding" from the government over a three-year period for R&D. In addition to lower interest rates, the loan approval time is expected to be as short as 20 minutes (Baker, 2020).

As of December 2019, 38 banks participated in the platform (Xinhua, 2020), which reached 48 by September 2020 (Ledger Insights, 2020). *During this period, the business volume* processed by the platform increased from US$12.4 billion (Xinhua, 2020) to US$29 billion (Ledger Insights, 2020). By August 2020, it had processed more than 50,000 transactions (Ledger Insights, 2020).

The BATFB's jurisdiction of operation is no longer restricted to the Bay Area. New regions are being added regularly. In addition to trade finance, it covers supply chain accounts receivables and rediscounting[1] by the central bank. Additional services include automated tax filing and supervision of international trades. The Institute is finding ways to expand the ecosystem. Especially it is interested in third parties that can build applications to integrate with the platform (Ledger Insights, 2020).

The auto manufacturer BYD is one of the participants. The majority of BYD's suppliers are SMEs, which include 10,000–20,000 tier one suppliers and even higher numbers of tier two and three suppliers. Especially the latter group is reported to have a more significant funding gap. By sharing data between the banks, a firm cannot submit the same invoice to two banks for funding. If an invoice is authenticated by companies such as BYD, it provides credibility for the bank. Put differently, blockchain's high specificity (von Briel et al., 2018) eliminates the possibility of fraudulent invoices.

Demonstrating Firms' and Products' Compliance with Various Regulations, Standards and Other Requirements

Blockchain-based solutions can also help demonstrate developing world-based firms' and products' compliance with various regulations, standards, and other requirements. As noted above, the limited availability of *data about quality-related* information, a lower *level of quality consciousness* among entrepreneurs in the informal sector, and unfavorable quality outcomes hinder small firms' access to markets (Hodges, 2012). Blockchain, in combination with other technologies, can help measure, document, and communicate *quality-related* information as well as information about entrepreneurial firms (e.g., the gender of entrepreneurs and their sustainability practices). For instance, Denver, Colorado-based startup Bext360's coffee tracking projects combine satellite images, blockchain, and artificial intelligence (AI). Bext360's kiosks in Uganda evaluate coffee beans using its Bextmachines. A Bextmachine is a Coinstar-like device, which employs smart image recognition technology

---

[1] Rediscounting is a way of providing financing by a central bank to banks or other financial institutions. If a customer wants to borrow $100, the bank may ask them to sign a paper document (also referred to as a note) promising to repay $120 in one year. In this example, the bank is "discounting" the note because it is paying less than $120. The central bank could provide financing to the bank by "rediscounting" the note (e.g., by paying $110 for the note).

machine vision, artificial intelligence, IoT and blockchain to grade and track coffee beans. It takes a three-dimensional scan of each bean's outer fruit (Cadwalader, 2018). Bextmachines analyze farmers' coffee cherries and coffee parchment deposited at collection stations and sort them to assess quality. Farmers that supply bigger and riper cherries are paid more. Bext360's systems store data related to the time, date, and location of transactions and the amount of payment.

The Bextmachines link the output to cryptotokens, which represent the coffee's value. New tokens are automatically created when the product passes through the supply chain. The values of tokens increase at each successive stage of the supply chain (Moyee., 2018).

Using a mobile app, relevant parties can negotiate a fair price. Farmers get paid immediately via the app. The app also determines the identity of the person selling the products. Using Bext360's API, intermediaries such as wholesalers and retailers embed the technology into their websites, marketing, and point of sales (PoS). This level of transparency may not be possible without blockchain. In April 2018, the world's first blockchain-traced coffee tracked by Bex360's solutions was sold (bext360, 2018). It also includes indicators related to sustainable sourcing and satellite images to show if producers are polluting water (Zhong, 2019). The combination of different technologies can help to get closer to the truth regarding sustainability practices.

The above measures are likely to address trust issues in the *gate for data before they enter* the blockchain network. Blockchain-based information's high specificity (von Briel et al., 2018) means that firms' and products' compliance with various regulations, standards, and other requirements can be verified with a high degree of certainty.

Efficient and Cost-Effective Way to Raise Capital

We noted above that traditional banks are unwilling and reluctant to serve poor people and small businesses due to high transaction costs and inefficient processes associated with making small loans (Adams & Nehman 1979). Cryptocurrencies could be an efficient and cost-effective way to raise capital to start a business. Brazil-based Moeda is an example of a startup that has made an attempt to use blockchain to promote entrepreneurship. Moeda describes itself as a blockchain-based microfinance platform that links investors with cooperative businesses (Schiller, 2018a, 2018b). The company's stated goal is to provide opportunities for under-banked individuals to establish digital identities and credit profiles, prove creditworthiness and build a reputation.

As noted above, ICOs are emerging as a key fundraising tool. This tool is also being utilized by entrepreneurs in developing economies. In 2017, Moeda raised US$20 million through an initial coin offering (ICO). There were about 2,000 token investors. US$10 million of that amount was allocated to funding projects (Schiller, 2018a, 2018b). Most of the investors for the ICO were young people from China. Foreign investors used blockchain tokenization to invest in Brazil.

Users can get micro-loans to start or expand businesses. Borrowers can use Moeda's peer-to-peer (P2P) payment app to pay for the things they need for their businesses. Moeda app can be used without any knowledge of cryptocurrency. Moeda works with cooperatives and banks to provide micro-loans to underserved populations (Radocchia, 2018). The borrowers are issued pre-paid debit cards, which allow monitoring of spending (Schiller, 2018a, 2018b). Entrepreneurs who receive loans pay interest rates of 15% with Moeda, which is significantly less than about 120% from other lenders.

Moeda's cooperative agriculture model allows it to capitalize on a community of local entrepreneurs. Moeda partnered with Unicafes, which has 100,000 members mostly small farmers in rural areas. The organization has stated that its aim has been to develop systems that allow investors to track the project's progress and accountability in a transparent manner (Field, 2018). Theoretically, if lenders want to invest in projects that have social impacts, they can (Isa, 2017).

Investors can buy Moeda's cryptocurrency MDA tokens and allocate to businesses through an app. Brazil's cooperative banks exchange the tokens for local currency (Schiller, 2018a, 2018b). This model helps to minimize costs associated with currency change, regulations, and other obstacles (Schiller, 2018a, 2018b). Moeda can cut middlemen and increase operational efficiency.

MDA is on the Ethereum blockchain. MDA tokens are the standard ERC20 token. Note that ERC20 is a technical standard used for smart contracts. It keeps track of token owners. Tokens can be stored in various wallets that are compatible with Ethereum (Anwaar, 2018).

A borrower was Hope Valley, a cooperative pumpkin and yucca farm and food processor in located in Formosa, Brazil. Moeda reported that Hope Valley received a six-month US$55,000 loan. The money was used to pay for an irrigation system and food processing equipment. It was reported that the co-op increased its production by fivefold (Field, 2018).

Blockchain-based solutions can make P2P lending possible by directly connecting lenders and borrowers thereby eliminating the need for intermediaries. As an example, consider the online lending platform Kiva. The company does not make direct loans. While some investors mistakenly think that Kiva offers direct person-to-person connections, it actually works with local MFIs as middlemen. Kiva says that it conducts audits of its Field Partners to ensure that poor people are not exploited. However, due to high overhead costs and other sources of inefficiency, Kiva Field Partners charge exorbitantly high-interest rates. For instance, a Kiva Field Partner in Senegal was reported to charge an interest rate of 40% (http://www.femalefounderstories.com/julia-kurnia.html).

Such loans could be made more affordable by eliminating the middlemen such as Kiva Field Partners. In this regard, programs such as Kiva's blockchain-based IDs are a first step toward improving access to finance for the poor. True decentralization will be complete when impact investors and philanthropic funders can directly reach the poor with cryptocurrencies.

Kiva worked with the U.N. Capital Development Fund (UNCDF) and the U.N. Development Program (UNDP) to develop a blockchain-based ID system in Sierra Leone. In August 2019, the government of Sierra Leone launched a blockchain-based National Digital Identity Platform (NDIP). Sierra Leone's government wants all banks and microfinance institutions (MFIs) to use the system. In Kiva's system, a borrower will be assigned a digital wallet, which will be accessible through an app. When a lender provides a loan, the borrower gets a verifiable claim with all details. The borrower accepts the claim. The loan is then posted to the borrower's private credit ledger in the Kiva wallet. The same process is repeated when the borrower makes a loan repayment. The borrower approves a verifiable claim sent by the lender, which is then posted to the ledger. All credit-related events are thus captured in a single ledger connected to the wallet controlled by an individual. Financial institutions, government agencies, and third-party agencies can access the information only with the owner's consent.

Sierra Leone has one credit bureau that has information only on the country's 2,000 people (https://www.kiva.org/blog/kiva-becomes-a-founding-partner-of-libra-association-to-support-financial-inclusion-for-all). Over 75% of Sierra Leone's population lacks access to formal banking services. People rely on informal institutions such as community banks and MFIs for their financial needs, which charge extremely high interest rates. They also do not share credit information.

As noted above, despite their high degree of relationality (von Briel et al., 2018), relationships created by some ICT applications such as social media are often shallow. Blockchain-based solutions create a higher degree of certainty in the information, which would increase the perceptions of integrity and hence trust required for coordination and cooperation (Pirson & Malhotra, 2011).

Overcoming Geopolitical Disadvantages in Accessing Key Resources and Handling Business Transactions

As noted above, due to their peripherality, developing world-based entrepreneurial firms face heightened challenges to access essential services and resources such as finance and the market (Anderson, 2000; Goodall, 1987). Developing world-based entrepreneurs face significant geopolitical disadvantages associated with this core–periphery relation in accessing key resources. For instance, the current global financial system is described as West-centric (Tiessen, 2015).

China's ambition has been to internationalize the Renminbi and reduce the dominance of the U.S. dollar in the global financial system. Belt and Road Initiative (BRI), which is a global development and industrial strategy adopted by the Chinese government in 2013, is emerging as a key force that can shape the current global financial system. Indeed, the BRI is viewed as the most ambitious infrastructure projects. The BRI has promised over US$1 trillion in infrastructure and covers more than 60 countries that account for two-thirds of the world's population (Perlez & Huang, 2017). China has launched a new international payments system, the China international payments system (CIPS). The first phase of the CIPS was launched in October 2015. Businesses in Asia and Europe can send funds in RMB to bank accounts in China, via CIPS without using the society for worldwide interbank financial telecommunication (SWIFT) system (Ding, 2017). The CIPS is especially attractive for countries that are adversely affected by U.S. sanctions. Iran has joined CIPS (Gasper, 2019). The Venezuelan government was also reported to be discussing adopting the CIPS (Argus Media, 2019). African nations that are receiving investments from China-led infrastructure projects under the

BRI are also using the CIPS. As of mid-2019, hundreds of banks from 89 countries had participated in this initiative.

In 2018, 2% of international transactions cleared on SWIFT were settled in Renminbi, which reduced to 1.4% in September 2020. As of 2020, CIPS was about 0.3% of the size of SWIFT. Efforts such as the BRI and CIPS are attempting to make the global financial system more China-centric (Joshua, 2019).

Many firms in developing countries face challenges in accessing credit in foreign currency. Due to import substitution policies in most developing countries, there are strict regulations on imports and foreign currency. Foreign currency is accessible only through the state (Pedersen & McCormick, 1999). Even if there are no such restrictions in a given country, small firms, especially in the informal sector face difficulty to access foreign currencies. Prior research suggests that firms that are less opaque with easily verifiable collateral, high net worth, reliant on formal financing rather than informal financing are more likely to get dollar credit (Mora & Neaime, 2013).

Due to their decentralized structure, blockchain and cryptocurrencies have no core–periphery structure. *For instance, Bitcoin* is free software and software *developers* anywhere in the world can contribute to the project. Small firms in the GS experience less hassle using decentralized blockchain-based solutions. Unsurprisingly crypto-denominated international commerce has become increasingly common among developing world-based small businesses (Orcutt, 2020). Small businesses in developing countries have reported that speed and efficiency can be greatly improved by making payments in cryptocurrencies rather than in major international currencies such as U.S. Dollar and Euro. A Nigerian vendor of *handsets and accessories, who* sourced his products from China and the United Arab Emirates reported that his Chinese suppliers prefer payments in cryptocurrency. He started paying with cryptocurrencies, which increased his profits. This is mainly because he did not have to buy U.S. dollars using the Nigerian naira or pay expensive fees to money-transfer agencies.

Due to such practical uses, which are unique to developing economies such as those in Africa, Bitcoin's use has been reported to grow in these economies. For instance, according to U.S. blockchain research firm Chainalysis, monthly cryptocurrency transfers of under US$10,000, which are typically made by individuals and small businesses, to and from Africa increased by more than 55% during June 2019-June 2020 to reach US$316 million (Akwagyiram & Wilson, 2020).

Chainalysis' analysis showed a similar pattern in Latin America. During June 2019–June 2020, Latin America sent $25 billion worth of cryptocurrency and received $24 billion (Orcutt, 2020). Chainalysis' data showed that East Asia was Latin America's significant counterparty (Chainalysis, 2020). The blockchain research firm's interviews with Latin America-based cryptocurrency operators indicated that many of the payments were commercial transactions between East Asia-based exporters and Latin American importers. A Paraguay-based cryptocurrency exchange explained that businesses in Paraguay import significant amount of goods from China. Some of them are then exported to other countries such as Brazil. Many of importers make payments using Bitcoin because of the speed and ease with which they can settle the payments. Due to concerns related to money laundering, banks in Paraguay are reluctant to do business with most companies. The banking application process is complex, which requires a many supporting documents and takes a long time. Even if a business's application to make payment in international currencies is approved, wire transfers are costly. Moreover, by making payments in cryptocurrencies, they can avoid import taxes (Chainalysis, 2020).

Some have developed innovative approaches based on blockchain that help employees receive *faster payment from their* employers. The Kenya-based blockchain startup BitPesa is helping to speed up the flow of cash from businesses in China to their African employees, who send money to their families. Launched in 2013, BitPesa uses Bitcoin to facilitate low-cost and instant payments (Gharib, 2017). As of the end of 2016, BitPesa operated in Kenya, Nigeria, Uganda, and Tanzania (Dyani, 2016).

In terms of relationality (von Briel et al., 2018), blockchain and cryptocurrencies can help to develop world-based entrepreneurs create a direct P2P relationship without trading partners. They are not required to rely on actors at the core (e.g., the West for making payments in US$ and Euro and China for Renminbi) under the current core–periphery model (Anderson, 2000; Goodall, 1987).

3.4   CHAPTER SUMMARY AND CONCLUSION

Blockchain makes it possible to help members of disadvantaged groups to engage in entrepreneurial activities. Blockchain-based solutions can also facilitate the operations of entrepreneurial firms in developing countries. The above examples, such as Circulor's plan to improve its machine learning models to distinguish between children and adults using aerial imagery and Oracle's partnership with the World Bee Project to monitor farmlands regarding their support to bee colonies and other pollinators indicate the potential of blockchain solutions to help entrepreneurial firms in the Global South to demonstrate that they protect workers from poor labor conditions such as child labor, forced labor, and bad working conditions and that they are engaged in ecologically responsible farming practices. For the population that lacks any digital footprint, initiatives such as those implemented by Kiva and BanQu might be the only way to make relevant information available to potential lenders.

Blockchain's potential to stimulate entrepreneurial activities among disadvantaged groups has not been fully realized. Many blockchain-based systems are designed to benefit big companies rather than disadvantaged groups. Due partly to this, despite the technical potential of blockchain to help disadvantaged groups, the benefits so far have been far smaller. Public policy measures are likely to play important roles in ensuring that disadvantaged groups can utilize information from blockchain-based transactions to engage in entrepreneurial activities.

REFERENCES

Adams, D. W., & Nehman, I. G. (1979). Borrowing costs and the demand for rural credit. *The Journal of Development Studies, 15*(2), 165–176.

Akwagyiram, A., & Wilson, T. (2020). *How bitcoin met the real world in Africa*. Reuters. https://www.reuters.com/article/us-crypto-currencies-afr ica-insight/how-bitcoin-met-the-real-world-in-africa-idUSKBN25Z0Q8

Anderson, A. R. (2000). Paradox in the periphery: An entrepreneurial reconstruction? *Entrepreneurship & Regional Development, 12*(2), 91–109. https://doi.org/10.1080/089856200283027

Anwaar, A. (2018). *Moeda: Humanizing finances, distributing impact*. Block Publisher. https://blockpublisher.com/moeda-humanizing-finances-distribut ing-impact/

Argus Media. (2019, July 17). *Caracas exploring Chinese, Russian payment systems*. Argus Media. https://www.argusmedia.com/en/news/1941803-car acas-exploring-chinese-russian-payment-systems

Baker, P. (2020). *China Injects $4.7M Into central bank's blockchain trade finance platform*. Coindesk. https://www.coindesk.com/china-injects-4-7m-into-cen tral-banks-blockchain-trade-finance-platform

Bardasi, E., Blackden, M. C., & Guzman, J. C. (2007). *Gender, entrepreneurship, and competitiveness in Africa*. Africa Competitiveness Report, Chapter 1.4. The World Bank.

Bbun, T. M., & Thornton, A. (2013). A level playing field? Improving market availability and access for small scale producers in Johannesburg, South Africa. *Applied Geography, 36*, 40–48.

Beck, T., & Cull, R. (2014). SME finance in Africa. *Journal of African Economies, 23*, 583–613.

bext360. (2018). *bext360 and coda coffee release the world's first blockchain-traced coffee from bean to cup*. Globenewswire. https://globenewswire.com/news-release/2018/04/16/1472230/0/en/bext360-and-Coda-Coffee-Release-The-World-s-First-Blockchain-traced-Coffee-from-Bean-to-Cup.html

Block, J. H., Colombo, M. G., Cumming, D. J., & Vismara, S. (2018). New players in entrepreneurial finance and why they are there. *Small Business Economics, 50*(2), 239–250.

Cadwalader, Z. (2018, April 23). *Trace your coffee using blockchain*. https://spr udge.com/132380-132380.html

Chainalysis. (2020). *How Latin America mitigates economic turbulence with cryptocurrency*. Cahinalysis. https://blog.chainalysis.com/reports/latin-america-cryptocurrency-market-2020

Chanjaroen, C., & Boey, D. (2016). *Fraud in $4 Trillion trade finance has banks turning digital*. http://www.bloomberg.com/news/articles/2016-05-22/fraud-in-4-trillion-trade-finance-turns-banks-to-digital-ledger.

Charness, J. (2019). *How oracle and the world bee project are using ai to save bees*. Oracle AI and Data Science Blog. https://blogs.oracle.com/datascience/how-oracle-and-the-world-bee-project-are-using-ai-to-save-bees-v2

Crypto Confidential. (2020). *Bitcoin's guardian angel; The 50 biggest companies in blockchain*. Forbes. https://www.forbes.com/sites/cryptoconfidential/2020/02/23/bitcoins-guardian-angel-the-50-biggest-companies-in-blockc hain/#25a0ed083fcf

Cumming, G. S., Cumming, D. H. M., & Redman, C. L. (2006). Scale mismatches in social-ecological systems: Causes, consequences, and solutions. *Ecology and Society, 11*(1), 14. http://www.ecologyandsociety.org/vol11/iss1/art14/

Cumming, D., Hou, W., & Lee, E. (2016). Sustainable and ethical entrepreneurship, corporate finance and governance, and institutional reform in China. *Journal of Business Ethics, 134*(4), 505–508.

Das, S. (2017). *Chinese fintech firms launch blockchain supply chain finance platform*. Cryptocoinsnews. https://www.cryptocoinsnews.com/chinese-fintech-firms-launch-blockchain-supply-chain-finance-platform/

DeSanctis, G., & Poole, M. (1994). Capturing the complexity in advanced technology use: Adaptive structuration theory. *Organization Science, 5*(2), 121–147.

Ding, Q. (2017, October 19). China breakthroughs: New international payments system goes into action. CCTV. http://english.cctv.com/2017/10/19/ART IXz5AbEqSewRu8jDIFRKO171019.shtml

Dyani, Z. (2016). *Op Ed: Africa needs more bitcoin and blockchain education*. Bitcoin Magazine. https://bitcoinmagazine.com/articles/op-ed-africa-needs-more-bitcoin-and-blockchain-education-1482249305/

Faridi, O. (2020). *People's bank of China Acquires $4.7 Million in funding to further develop blockchain-based trade finance platform*. Crowdfund Insider. https://www.crowdfundinsider.com/2020/03/158535-peoples-bank-of-china-acquires-4-7-million-in-funding-to-further-develop-blockchain-based-trade-finance-platform/

Faulkner, J., & Runde, P. (2009). On the identity of technological objects and users' innovation in functions. *Academy of Management Review, 34*(3), 442–462.

Field, A. (2018). *A bitcoin for women entrepreneurs in Brazil and impact investors globally*. Forbes. https://www.forbes.com/sites/annefield/2018/01/24/a-bitcoin-for-women-entrepreneurs-in-brazil-and-impact-investors-glo bally/#674e05936cee

finextra.com. (2019). *Banks back pilot bidding to unlock finance for sustainability in supply chains*. Finextra. https://www.finextra.com/newsarticle/34401/banks-back-pilot-bidding-to-unlock-finance-for-sustainability-in-supply-chains

Fombang, M. S., & Adjasi, C. K. (2018). Access to finance and firm innovation. *Journal of Financial Economic Policy, 10*(1), 73–94. https://doi.org/10.1108/JFEP-10-2016-0070

Gasper, D. (2019, June 16). *Pain of tariffs and sanctions behind China and Russia's push to dethrone the US dollar*, South China Morning Post. https://www.scmp.com/comment/opinion/article/3014258/pain-tar iffs-and-sanctions-behind-china-and-russias-push-dethrone

Gharib, M. (2017). *Blockchain could be a force for good. But first you have to understand it*. NPR. http://www.npr.org/sections/goatsandsoda/2017/01/11/503159694/blockchain-could-be-a-force-for-good-but-first-you-have-to-understand-it

Giddens, A. (1984). *The constitution of society*. Polity Press.

Goodall, B. (1987). *The dictionary of human geography*. Penguin.

Gould, E. (2016). *Africa's big banks are betting on fintech startups and bitcoin to beat disruption*. https://qz.com/618674/africas-big-banks-are-betting-on-fintech-startups-and-bitcoin-to-beat-disruption/

Hanstad, T. (2013). *The case for land reform in India*. https://www.foreignaffairs.com/articles/

Hodges, R. J. (2012, September). Natural Resources Institute University of Greenwich Postharvest Quality Losses of Cereal Grains in Sub-Saharan Africa. http://archive.aphlis.net/downloads/Postharvest_quality_losses%20review_Sept_2012_revised.pdf

Isa. (2017). *New Year special: All you want to know about MOEDA*. Medium. https://medium.com/moeda/new-year-special-all-you-want-to-know-about-moeda-703e056d825c

Joshua J. (2019). International financial organizations and the BRI. In *The belt and road initiative and the global economy*. Palgrave Macmillan, Cham. https://doi.org/10.1007/978-3-030-28030-7_7

Kallinikos, J., Aaltonen, A., & Marton, A. (2013). The ambivalent ontology of digital artifacts. *MIS Quarterly, 37*, 357–370.

Kshetri, N. (2016). Fostering startup ecosystems in India. *Asian Research Policy, 7*(1), 94–103.

Kshetri, N. (2019). Global entrepreneurship: Environment and strategy.

Kshetri, N. (2017). Will blockchain emerge as a tool to break the poverty chain in the Global South? *Third World Quarterly, 38*(8), 1710–1732.

Kshetri, N. (2020). Blockchain-based financial technologies and cryptocurrencies for low-income people: Technical potential versus practical reality. *IEEE Computer, 53*(1), 18–22.

Kshetri, N. (2021). *Blockchain and supply chain management*. Elsevier.

Kshetri, N. (2023). The nature and sources of international variation in formal institutions related to initial coin offerings: Preliminary findings and a research agenda. *Financial Innovation, 9*(1), 9. https://doi.org/10.1186/s40854-022-00405-x

Ledger Insights. (2020). China wants to link central bank trade finance blockchain with Asia, Europe. *Ledger Insights*. https://ledgerinsights.com/china-central-bank-trade-finance-blockchain-asia-europe/

Macdonald, K. (2007). Globalising justice within coffee supply chains? Fair trade, starbucks and the transformation of supply chain governance. *Third World Quarterly, 28*(4), 793–812.

Mallalieu, K., & Mark, L. (2012, March 21). Mobile apps boost Trinidad and Tobago fish market, *Digital Opportunity*.

Momta, P. P. (2020). Entrepreneurial finance and moral hazard: Evidence from token offerings. *Journal of Business Venturing*, 106001. https://doi.org/10.1016/j.jbusvent.2020.106001

Mora, N., & Neaime, S. (2013). Foreign currency borrowing by small firms in emerging markets: When domestic banks intermediate dollars. *Journal of Banking & Finance, 37*, 1093–1107.

Morris, N. (2018). *MonetaGo's blockchain solution for trade finance fraud.* https://www.ledgerinsights.com/monetago-blockchain-trade-finance-fraud/

Moyee. (2018). *World's first blockchain coffee project.* Moyee Coffee. https://moyeecoffee.ie/blogs/moyee/world-s-first-blockchain-coffee-project

Mwai, C. (2018). Will blockchain fix the mineral traceability woes? *The New Times.* https://www.newtimes.co.rw/news/will-blockchain-mineral-tra ceability

Nambisan, S. (2017). Digital entrepreneurship: Toward a digital technology perspective of entrepreneurship. *Entrepreneurship Theory and Practice, 41*(6), 1029–1055.

O'Neal, S. (2019, July 29). *Differences between tokens, coins and virtual currencies, explained.*

Orcutt, M. (2020). *Cryptocurrency may be supercharging trade between Latin America and Eastern Asia.* The Block. https://www.theblockcrypto.com/post/76839/cryptocurrency-eastern-asia-latin-america-trade-chainalysis

Orlikowski, W. J., & Scott, S. V. (2008). Challenging the separation of technology, work and organization. *The Academy of Management Annals, 2*(1), 433–474.

Ortmann, G. F., & King, R. P. (2007). Agricultural cooperatives II: Can they facilitate access of small-scale farmers in South Africa to input and product markets? *Agrekon, 46*(2), 219–244. https://doi.org/10.1080/03031853.2007.9523769

Pedersen, P. O., & McCormick, D. (1999). African business systems in a globalising world. *The Journal of Modern African Studies, 37*(1), 109–135.

Pergelova, A., Manolova, T., Simeonova-Ganeva, R., & Yordanova, D. (2019). Democratizing entrepreneurship? Digital technologies and the internationalization of female-led SMEs, *Journal of Small Business Management, 57*(1), 14–39. https://doi.org/10.1111/jsbm.12494

Perlez, J., & Huang, Y. (2017). Behind China's $1 Trillion plan to shake up the economic order. *The New York Times.* https://www.nytimes.com/2017/05/13/business/china-railway-one-belt-one-road-1-trillion-plan.html

Peterson, K. (2019). *Foxconn founder: Libra can 'Converge' with China's digital currency in Taiwan.* Moon Catcher. https://dailynews.bitcoindiamond.org/foxconn-founder-libra-can-converge-with-chinas-digital-currency-in-taiwan/

Pirson, M. A., & Malhotra, D. (2011). Foundations of organizational trust: What matters to different stakeholders? *Organization Science, 22*(4), 1087–1104.

Pitti, A. (2018). *Why India can become the global center for blockchain innovation.* Nasdaq. https://www.nasdaq.com/article/why-india-can-become-the-global-center-for-blockchain-innovation-cm992358

Porter, M. E., & Kramer, M. R. (2002). The competitive advantage of corporate philanthropy. *Harvard Business Review, 80*(12), 56–68.

Radocchia, S. (2018). *Why controlling your credit history is possible through blockchain technology*. Medium. https://keepingstock.net/why-contro lling-your-credit-history-is-possible-through-blockchain-technology-b2778d eed391

Ramirez, V. (2017, June 11). *How blockchain can make identification borderless and immutable*. Singularity Hub. https://singularityhub.com/2017/06/11/how-blockchain-is-helping-in-the-battle-against-extreme-poverty/#sm.001dojtf51cf3dbxr292hdqg31bks

Ross, S. A. (1973). The economic theory of agency: The principal's problem. *American Economic Review, 63*(2), 134–139.

Schiller, B. (2018a). *Can the world's underbanked leapfrog on to a new blockchain financial system?* Fast Company. https://www.fastcompany.com/40524015/can-the-worlds-underbanked-leapfrog-on-to-a-new-blockchain-fin ancial-system

Schiller, B. (2018b). *Can the world's underbanked leapfrog on to a new blockchain financial system*. Fast Company. https://www.fastcompany.com/40524015/can-the-worlds-underbanked-leapfrog-on-to-a-new-blockchain-fin ancial-system

Stanley, A. (2017, December 8). *Microlending startups look to blockchain for loans*. Coindesk. https://www.coindesk.com/microlending-trends-startups-look-blockchain-loans

Stiglitz, J. E., & Weiss, A. (1981). Credit rationing in markets with imperfect information. *American Economic Review, 71*(3), 393–410.

Suberg, W. (2017). *Indian Bank wants joint effort to share data on blockchain*. The Coin Telegraph. https://cointelegraph.com/news/indian-bank-wants-joint-effort-to-sha///re-data-on-blockchain

Sustainable Food News. (2018). *USDA warns industry of 6 fake organic certificates*. Sustainable Food News. https://sustainablefoodnews.com/usda-warns-industry-of-6-fake-organic-certificates/

Tiessen, M. (2015). The appetites of app-based finance. *Cultural Studies, 29*(5–6), 869–886. https://doi.org/10.1080/09502386.2015.1017148

Van der Meer, C. L. J. (2006). Exclusion of small-scale farmers from coordinated supply chains: Market failure, policy failure or just economies of scale? In R. Ruben, M. Slingerland, & H. Nijhoff (Eds.), *Agro-food chains and networks for development* (Vol. 14, pp. 209–217). Wageningen UR Frontis Series, Springer.

von Briel, F., Davidsson, P., & Recker, J. C. (2018). Digital technologies as external enablers of new venture creation in the IT hardware sector. *Entrepreneurship Theory and Practice, 42*(1), 47–69.

WEF [World Economic Forum]. (2015). *Deep shift technology tipping points and societal impact survey report.* http://www3.weforum.org/docs/WEF_GAC15_Technological_Tipping_Points_report_2015.pdf

West, Darrell M. (2012). *How mobile technology is driving global entrepreneurship.* Brookings Institution. http://www.insidepolitics.org/brookingsreports/m_entrepreneurship.pdf

Williamson, O. E. (2002). The theory of the firm as governance structure: From choice to contract. *Journal of Economic Perspectives, 16*(3), 171–195.

Xinhua (2020). *China sees more banks using blockchain platform: Central Bank.* Xinhuanet. http://www.xinhuanet.com/english/2020-02/21/c_138806484.htm

Zahra, S. A., Priem, R. L., & Rasheed, A. A. (2007). Understanding the causes and effects of top management fraud. *Organizational Dynamics, 36*(2), 122–139.

Zetsche, D., Buckley, R., Arner, D., & Barberis, J. (2018). From fintech to techfin: The regulatory challenges of data-driven finance. *New York University Journal of Law & Business, 14*, 393–446.

Zhong, C. (2019, August 14). *Innovator BanQu builds blockchain and bridges for traceability, small farmers' livelihoods.* Greenbiz. https://www.greenbiz.com/article/innovator-banqu-builds-blockchain-and-bridges-traceability-small-farmers-livelihoods

Zucker, L. G. (1988). Where do institutional patterns come from? Organizations as actors in social systems. In L. G. Zucker (Ed.), *Institutional Patterns and Organizations: Culture and Environment* (pp. 23–49). Ballinger.

CHAPTER 4

Blockchains with Chinese Characteristics

Abstract The Chinese blockchain industry and market have unique, unusual, and idiosyncratic characteristics. Several established companies as well as start-ups in China have developed and deployed innovative solutions involving blockchain. On the other hand, some uses of blockchain such as initial coin offerings for capital-raising have been banned. While blockchain is a disruptive technology with the potential to transform the economy and society, it promotes decentralized governance and thus can turn centralized political structures such as those of China upside down. This chapter offers an analysis of seemingly conflicting actions of the Chinese government on the blockchain front. It analyzes how China is creating a delicate balance between encouraging the use of blockchain for the growth and modernization of the economy and making sure that the technology is not used to threaten the legitimacy of the Chinese Communist Party.

Keywords 3TG minerals · China · Chinese Communist Party · Initial coin offering · Techno-utilitarian model · The EU Conflict Minerals Regulation

4.1   INTRODUCTION

China has become a global epicenter of blockchain-related activities. As early as 2018, the country had more new blockchain companies than in the U.S. (Vidrih, 2018). According to blockchain and crypto data platform LongHash, as of August 2020, there were more than 84,410 registered blockchain companies in China and 29,340 of them were in operation. About 10,000 of them were registered in the first seven months of 2020 (Peng, 2020).

Some important practical uses of blockchain have been reported. For instance, during the COVID-19 pandemic, blockchain-based solutions allowed 17 million people to travel between Guangdong and Macau by mid-October 2020 (Feng, 2020). The mainland China-Macau blockchain health code system was developed by the Chinese open-sourced blockchain platform FISCO BCOS and the technology company Tencent's neobank WeBank. A consortium blockchain network was used, in which data can be accessed only by authorized organizations. The system uses blockchain to encrypt the identification and personal health information of travelers. Health authorities in mainland China and Macau need to verify the health information submitted by users that cross the border. However, the regulations of the two jurisdictions do not allow the exchange of personal data directly with each other.

Blockchain-based solutions have also improved food safety and quality. *Among many uses of* blockchain in this industry, the most common one is for stopping fake and counterfeit products. For instance, blockchain solutions are being used to stop counterfeit versions of the Wuchang rice, which is known for high quality (Sunny, 2018) and dairy and meat products imported from New Zealand and Australia (Hsu, 2018).

The Chinese *government, however, has been strongly against* some types and uses of blockchain. For instance, in September 2017, China banned the blockchain-based capital-raising method known as initial coin offering (ICO), also referred to as a *token offering*. About 90 cryptocurrency exchanges and 85 ICOs were shut down that year. During that year, the world's total bitcoin traded in yuan also decreased from 90% to less than 5% (Marinoff, 2018). In general, the Chinese Communist Party (CCP) views more decentralized blockchains in a negative way. It prefers blockchains that it can access and control. In decentralized blockchains, the founders, development resources, and crypto-assets are out of the CCP's reach and are thus undesirable (McClurg & Talati, 2018).

In order to understand the opportunity that blockchain provides, the way that the modern economy functions must be clear. One way to view a modern market economy is as trusted relationships that are maintained with various "ledgers" (Berg et al., 2018). For instance, ledgers maintained by governments consist of information about various aspects such as citizenship, tax obligations of businesses and citizens, social security and property ownership, passports, natural resources such as forests, land use and forms of land, and intellectual property. Likewise, businesses' ledgers have information about their employment, assets, supply chain partners, and customers (Pearson, 2019). In the non-blockchain world, these ledgers and records are *maintained by centralized* entities such as governments and firms (Abadi & Brunnermeier, 2018). Blockchain turns centralized management of records upside down by enabling decentralized governance of such records (Allen et al., 2019). Blockchain thus is an institutional governance mechanism for creating and maintaining distributed ledgers of information (Berg, 2017).

These opportunities are also accompanied by a number of challenges. For instance, centralized entities no longer have complete control over records. Such challenges are especially significant for authoritarian regimes, which tend to control and monopolize information flow within their border.

4.2 Key Features of the Chinese Blockchain Industry and Market

The Chinese blockchain industry and market have unusual and idiosyncratic characteristics. It can be attributed to the fact that today's China is a unique political structure. China has been described as a combination of an empire and a modern nation (Terrill, 2005). In general, China's state strategies toward ICTs have been to balance economic modernization and political control (Kalathil, 2003). Regarding political control, Reporters without Borders noted that "China was one of the first countries to realize it couldn't do without the Internet and so it had to be brought under control" (McLaughlin, 2005). China thus focused its attention on the Internet before most other developing countries in order to maintain control (Yang, 2001). This vision has broadly shaped the development of the Chinese blockchain industry and market.

One *observation that is worth* mentioning is that there have been instances of blockchain use in China that have been problematic from the point of view of political control. Thanks to blockchain, Chinese Internet users achieved some victories in the fight against the country's strict Internet censorship. A historic moment was made on April 23, 2018. Peking University's former student, Yue Xin had written a letter detailing the university's attempts to hide sexual misconduct. The case involved a student, Gao Yan, who committed suicide in 1998 after a professor sexually assaulted and then harassed her. The letter was blocked by Chinese social networking websites, but an anonymous user posted it on the Ethereum blockchain (Kshetri, 2019).

In another case, in July 2018, Chinese citizens used blockchain to preserve an investigative story that condemned inferior vaccines being given to Chinese babies. The vaccines produced by Shenzhen-based Changsheng Bio-Tech failed to fight tetanus and whooping cough. The company has also allegedly faked data for about 113,000 doses of human rabies vaccine.

Yue Xin's letter, which was written in English and Chinese, and the story about the inferior vaccines have been inserted into the metadata of transactions in the Ethereum blockchain. Each transaction cost a few cents. Since Ethereum transactions are permanent and public, anyone can read the letter. The posts cannot be tampered with. Since they are distributed among many computers in decentralized networks, it is not possible for Chinese Internet censors to pressure any company to remove them (Kshetri, 2019).

The Techno-Utilitarian Model

Among key aspects that underpin the advantage of blockchain concern is its ability to protect privacy and strengthen security (Kshetri, 2017). In this regard, it is worth noting that China's approach to data privacy and security has been described as "techno-utilitarian." The utilitarian approach is a fundamental ethics principle where a decision is made based upon the greatest good for the greatest number of people. Thus, the protection of personal privacy and individual rights receives less emphasis. Some critics have noted that such arguments have been used to justify the use of AI for surveillance purpose and to suppress minorities and political opponents such as Muslims in the Xinjiang province (Thornhill, 2019). This broad pattern is precisely what one would expect in China's deployment of blockchain.

Some regulations have been laid down to control blockchains, which are directed toward addressing blockchain's censorship-resistance features. The "Regulations on the Management of Blockchain Information Services" became effective on February 15, 2019 (Kshetri, 2019). The regulation requires users to provide real names as well as national ID card numbers, mobile phones, or company registration to use blockchain services. User anonymity is thus not allowed. Blockchain services are required to remove "illegal information" quickly in order to stop it from spreading among users. Providers of blockchain services are also required to retain backups of user data for six months. Moreover, law enforcement must be able to get access to data whenever it is necessary (Liao, 2018). Blockchain service providers are required to keep relevant records of transactions and report to authorities in case of illegal use. They are also obliged to prevent the production, duplication, publication, and dissemination of contents that are banned by Chinese laws.

Unfavorable Environment for Certain Uses of Blockchain

Some of the key areas that are driving the use of blockchains in other major economies are less common in China. One such application domain is increasing supply chain traceability to ensure that serious human rights abuses such as child labor and forced labor do not take place. Since the 2010s, regulators in Western countries have started responding to serious human rights violations by their companies even if such violations take place outside their jurisdictions.

The Dodd–Frank Wall Street Reform and Consumer Protection Act requires U.S. companies to vet their supply chains in the Democratic Republic of the Congo and neighboring countries (Ayogu & Lewis, 2011). Section 1502 of the Act requires mining companies to disclose if they source conflict minerals: tin, tungsten, tantalum, and gold—from DRC and nine neighboring countries (Mwai, 2018). The EU Conflict Minerals Regulation (EU *Regulation No. 2017/821*) was adopted in May 2017 by the EU Parliament and EU Council (European Union, 2017a). The new law is expected to come into force in the EU in 2021. From January 1, 2021, importers of tin, tantalum, tungsten, and gold in the EU will be required to carry out due diligence on their supply chains. That is, they need to check the sources of the minerals and metals they import and ensure that they were processed responsibly (European Union, 2017b). This import regulation covers *the so-called 3TG* minerals (tin, tantalum and tungsten, their ores, and gold). Companies importing these minerals will be impacted by the regulation. The regulation is expected to affect the activities of more than 1,000 importers directly and tens of thousands of economic actors will indirectly (Thomas, 2020). For instance, smelters, refiners. OEMs, component manufacturers, and other businesses that use 3TG to manufacture products are required to source 3TG responsibly (Hussey, 2020). The European Union's 2020 report "Study on due diligence requirements through the supply chain" has recommended that the use of technology such as blockchain be explored to enhance supply chain visibility and reduction of supply chain complexity (European Union, 2020). The 2017 French Duty of Vigilance Law has taken responsible sourcing one step further by imposing a mandatory due diligence requirement for human rights and negative environmental impacts (European Union, 2020). Likewise, starting in 2025, the London Metal Exchange (LME) plans to ban metals from its lists of approved brands if the extraction involves child labor and other abuses. In its initial plan announced in April 2019, the exchange had set 2022 as the deadline to comply with the guidelines. The deadline needed to be postponed because some major producers reportedly complained about the difficulties in meeting the requirements (Desai et al., 2019).

China lacks legislative commitment to responsible sourcing. Regarding the LME's plans to ban the trading of metals involving child labor and other abuses, Chinese firms' support is especially important since China is the world's largest consumer of industrial metals. However, what is referred to as the "China barrier" has been a major concern in the implementation of responsible sourcing initiatives in the global metal and mineral industry (Desai et al., 2019). Whereas industrialized countries in North America, Western Europe, Oceania, and other parts are strengthening regulatory and enforcement mechanisms to address the issues related to conflict minerals, China lacks such measures. Citing industry sources, news *website* Reuters stated that the absence of legislation and the lack of experience in sourcing conflict-free minerals have been major hindrances to China's engagement in the process (Desai et al., 2019). For instance, in China, there are no regulations equivalent to the Dodd–Frank Wall Street Reform and Consumer Protection Act of the U.S. or the Modern Slavery Act 2015 of the U.K.

Compatibility of Various Blockchain Types in China

China's political structure has a strong impact on the types of blockchain likely to be used in the country, especially in terms of consensus mechanisms. It is important to note that in a shared ledger, an efficient, fair, and secure mechanism is needed in order to make sure that only genuine transactions occur and participants agree on the ledger's status. A consensus mechanism performs this task by defining a set of rules to decide the various participants' contributions. The goal is to achieve the necessary agreement on a data value or the network's state (Frankenfield, 2020). There are three major types of consensus mechanisms: Proof of Work (POW), Proof of Stake (PoS), and Proof of Authority (PoA).

In a POW protocol, all users can compete to verify transactions. Major drawbacks of such protocol include high energy consumption and longer processing time. In a PoS consensus model, only a small group of nodes can validate transactions. A node's power to validate transactions or responsibility to maintain the public ledger is proportional to the number of virtual currency tokens associated with the node (Gazdecki, 2019). For instance, a node that owns 5% of the currency available theoretically can validate only 5% of the blocks. It is viewed as a low-cost and low-energy consuming alternative to the POW algorithm. Finally, a PoA consensus model relies on a limited number of trustworthy block validators, which are pre-approved. It is viewed as a modified form of PoS, in which a validator's identity rather than the role of stake is important. The nodes responsible for validating transactions are selected based on certain rules (POA Network, 2017).

What is clear is that the Chinese government is against truly decentralized blockchain systems such as the Bitcoin which use PoW protocols. Yao Qian, the head of the PBoC digital currency research institute argued that blockchain needs to be centralized (Wilmoth, 2018). He noted that in a multi-center system, in which consensus is managed by several main nodes, intervention can be applied in case of emergency, data can be *rolled back and transactions can be reversed*. Under the structure he proposed, the PBoC could shut down the system and reverse transactions and push software upgrades. There is no need for community consensus.

China has been the first nation to rank blockchains (Emsley, 2018). The country's Center for Information and Industry Development (CCID), under the Ministry of Industry and Information Technology has been regularly publishing its Global Public Blockchain Technology Assessment Index (GPBTAI). Most of the top-ranked blockchains in the CCID assessment are not top blockchain companies as measured by market capitalization. It published the 17th GPBTAI index in April 2020. The April 2020 index ranked EOS #1. Bitcoin ranked 14th on the list (Faridi, 2020).

EOS uses the delegated-proof-of-stake (DPoS) consensus model. The difference between DPoS and PoS can be compared to the difference between direct democracy and representative democracy. In a regular PoS protocol, every wallet containing coins is able to "stake." That is, it can validate transactions and form a distributed consensus. The wallet earns coins in return. In a DPoS system, a wallet with coins can vote for delegates. The delegates validate transactions and maintain the blockchain. They take the transaction fees as profit (cointelegraph, 2018).

In a DPoS model, users vote for representatives to make decisions for them. Only the representatives can verify transactions and make decisions regarding system updates. A flaw of the DPOS consensus model is that an average voter's chance of impacting the delegates that are selected is small. It is argued that a potential voter's incentive to vote based on the interest of the community is small. Voters are thus more likely to be motivated by bribes offered to them (Buterin, 2018).

While EOS claims that it is a decentralized, democratic network, all transactions and governance decisions are processed and approved by only 21 users, known as supernodes or block producers (BPs). In 2018, twelve of the supernodes were in China (Caijing, 2018). The fact that the majority of the supernodes are in China makes it easy for the Chinese government to control their activities.

Likewise, VeChain uses PoA consensus. Its 101 Authority Masternodes work as validators. Anonymous nodes are not allowed. Disclosure of identity is required to become an authority Masternode (Frankenfield, 2019).

4.3   Technological and Innovation-Related Activities in Blockchain

Many blockchain solutions have been launched by established companies as well as innovative start-ups. Among established companies, Alibaba, which holds more patents than any other Chinese firm, has launched many blockchain solutions (Table 4.1). In 2018, it launched food-focused blockchain platform, known as the Food Trust Framework, which provides traceability, visibility, and transparency in the supply chain. It is arguably China's first blockchain-based global product traceability program. Products sold in Alibaba's online retail system Tmall Global marketplace can be traced with blockchain and product tagging with unique QR codes (Taylor, 2018).

The Shanghai-based BaaS provider BitSE's VeChain is among the most prominent solutions launched by Chinese blockchain start-ups. VeChain combines blockchain and near field communication (NFC), RFID tag or QR code to fight counterfeiting. Note that NFC allows devices to communicate with each other when they are near. A customer's phone can communicate with the VeChain chip that is embedded inside the clothing or accessory. Customers can also scan QR codes on the label.

Table 4.1 Blockchain patents filed by Chinese firms

Study conducted by	Period covered	Findings
Tokyo-based research firm Astamuse	2009–2018	Total no. of patents worldwide 12,000. China: about 7,600, the U.S.: 2500. South Korea: 1,150, Japan: 380 China overtook the U.S. in patent submissions in 2016 (Hashimoto & Cho, 2019)
China Academy for Information and Communications Technology (CAICT)	2013–December 20, 2018	China applied for 4,435 patents (48% of the total), compared to 1,833 by the U.S. (21% of total (Ledger Insights, 2019)
Patents filed with the World Intellectual Property Organization (WIPO)	2017	Chinese firms accounted for 99 of the 314 patents (Peyton, 2019)
IPRDaily (media outlet focusing on IP), and incoPat (Global patent database)	2019	Chinese companies took the top three spots (Alibaba, Tencent and Ping An) well as seven among the top 10 and 19 among the top 30 (Chilnadaily.com.cn, 2020)

Each product is assigned a unique ID and VeChain stores the information on the blockchain. The information is also attributed to the product (e.g., with an NFC chip) (https://boxmining.com/vechain/).

In Focus 4.1: VeChain's Internationalization Initiatives

Chinese blockchain companies such as VeChain are already active in several foreign countries. As early as in 2017, VeChain solutions were used also in Europe and South East Asia (O'Connell, 2017). In November 2019, VeChain reached an agreement with logistics, supply chain and import solutions provider ASI Group to implement cross-continental logistics and trading solutions using VeChainThor blockchain (Brown, 2019). Together with the global quality assurance and risk management company *DNV GL*, VeChain announced Foodgates (Brown, 2019). The plan is to develop a platform to monitor products along the value chain (e.g., cow selection, slaughter, packaging and shipping to the supermarket/buyer for beef products).

Italian startup *Tokenfarm* aims to introduce emerging technologies such as blockchain, artificial intelligence, the Internet of Things, and 5G networks to the agrifood industry (Lucien, 2020). Tokenfarm, is an initiative of Confederazione Nazionale Coltivatori Diretti (Coldiretti), which is Italy's main farmers' organization that has over 1.6 million members. In July 2020, Coldiretti and food firm *Princes* signed a three-year contract to track the value chain of Italian tomatoes. every year, Princes plant processes 300 million kilograms of tomatoes pass involving 300 companies, 17 cooperatives and six associations of producers (Lucien, 2020). The blockchain pilot is powered by VeChain's ToolChain.

As of early 2020, VeChain had developed more than 40 enterprise applications on its platform (Garg, 2020). VeChain is a co-founder of the Belt and Road Initiative Blockchain Alliance (BRIBA) established in December 2019. The BRIBA aims to contribute to the Belt and Road Initiative (BRI) by leveraging blockchain (VeChain, 2019). Note that the BRI is a global development and industrial initiative proposed by the Chinese government in 2013.

Sensors are used to measure various indicators related to a product and record information at every stage of the supply chain. Customers can verify whether an item is genuine or not (Campbell, 2017). VeChain chip holds a unique public and a private key. The public key is stored in the blockchain, which can be verified by VeChain app. The app verifies the public key with the VeChain servers to know whether the public key is genuine (Campbell, 2016). At any point in the supply chain, distributors, retailers, or consumers can interact with the chip, tag or code. Doing this, companies can access to information about how the products are handled. The information is linked to the product's identity.

VeChain designs the sensors and manufacturers such as Bosch and Qualcomm produce them. As of May 2017, BitSE was reported to have more than two million product identities on VeChain (Reutzel, 2017).

Consumers can verify the products they bought are legitimate and handled appropriately. When consumers scan a QR code, they will be directed to a website that displays detailed information about the product such as its journey from the origin to the store. Each transaction has a timestamp. Information about the supplier and logs of temperature measurements in the supply chain are recorded. The QR code also has information about the blockchain transaction ID, which shows where all relevant data is stored on the VeChainThor blockchain (vechainin-sider.com, 2019). Additional information about appropriate ways to use and store the product can also be provided.

Using VeChain's ToolChain, companies can become ToolChain Admin Center and build systems to handle transactions on the VeChainThor blockchain. Companies can upload product descriptions, features, photos and videos. ToolChain Admin Center also manages which supply chain participants can write information on to the VeChainThor blockchain. A ToolChain Admin can predefine the events to share in the blockchain and control and assign what different companies can do. It is also possible to see the transaction generated by the product's journey (VePress, 2019).

ToolChain offers services such as product lifecycle management, supply chain process control, data deposit, and certification of data and process (Vechain, 2020). ToolChain is combined with cold-chain IoT[1] and different types of software and hardware sensors. All supply chain members upload logistics and manufacturing data in the VeChainThor blockchain. The products are tagged with secured QR codes. Customers can scan the QR code with their smartphones to see the product's full logistical history such as country of origin, packaging information, and storage temperature. Additional relevant information is included in pictures and videos.

---

[1] Cold-chain IoT involves processes and technologies that are used to transport temperature-sensitive products from origin to destination.

The above blockchain-related activities are associated with and facilitated by China's investments in intangible assets such as patents and research and development. China has consistently ranked No. 1 in the world in terms of blockchain patent filing (Table 4.1). While some argue that patents are irrelevant in blockchains, others argue that patents reflect technological or innovative activities and output, which are key in private and hybrid blockchains (Heasman, 2020). China's global dominance of blockchain patents has implications for blockchain-based solutions in the global market.

While Chinese firms have a tendency to file a lot of blockchain patents, they lag behind the U.S. in granted patents (Gkritsi, 2020). According to China Patent Protection Association, as of May 14, 2020, there were 3,924 blockchain patents granted worldwide. China's share was less than half that of the U.S. (Fig. 4.1).

The problem of *quantity versus quality of patents* may also have implications for innovative blockchain solutions likely to be developed in China. Research suggests that 10% of patents account for about 90% of total patent value. Alternative measures such as patent citation are viewed as more useful indicators. In this regard, *not a single one of the top 100* most cited patents during 2003–2019, came from China. It is also argued that China's leading AI companies such as Tencent, Alibaba, and Baidu

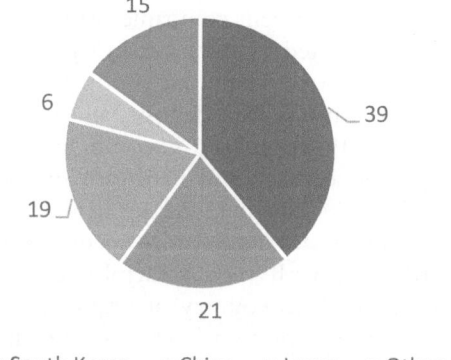

Fig. 4.1 Shares of blockchain patents granted worldwide (%) (*Data source* China Patent Protection Association [https://www.jinse.com/blockchain/747 938.html])

have just copied business models of U.S. companies such as Facebook, Amazon, and Google and tailored them to the Chinese market (Frey & Osborne, 2020).

4.4 KEY BLOCKCHAIN APPLICATIONS

In this section, we discuss key blockchain applications in China. In order to narrow down the blockchain use cases in China for this discussion, we start with the annual Blockchain 50 list published in February 2020 by Forbes. The list consists of the world's biggest brands with over $1 billion in annual revenue that are using blockchain. The list was first introduced in April 2019 (Castillo, 2019). An analysis of the Netherlands-based market intelligence platform for blockchain and DLT firm Blockdata found that companies in the Blockchain 50 were more likely to use blockchain for traceability and provenance, which are closely related to the supply chain, compared to payments and settlement (Fenton, 2020). Blockdata's analysis indicated that fifteen had used blockchain solutions in traceability and provenance, whereas thirteen had used such solutions for payments and settlements (Blockdata, 2020). Blockdata also found that six of the Blockchain 50 companies specifically developed supply chain management use cases. PwC has also identified provenance *as the No 1 use case* of blockchain and estimated that by helping organizations to verify the sources of their goods and track their movement and enhancing supply chain transparency, the technology has the potential to increase the global GDP by US$962 billion by 2030 (Pwc, 2020).

As noted above payments and settlements had the second-highest number of use cases among the Blockchain 50 companies. More broadly the financial technology (fintech) sector dominates blockchain patents worldwide with applications focusing on cryptocurrencies, their exchanges, and the use of blockchain to support financial transactions (*KissPatent*, UD). More than half of Chinese blockchain projects have focused on fintechs (Jao, 2019). A related area of application concerns insurance technology (InsurTech). In this section, we look at blockchain applications in China in two broad areas: (a) Improving *supply chain* traceability and provenance; (b) Cryptocurrency, FinTech, and InsurTech.

Improving Supply Chain Traceability and Provenance

Addressing Food and Drug Safety and Quality

Chinese as well as foreign companies have launched blockchain solutions for the Chinese market to track and trace products and enhance safety and quality in food and drug supply chains. Blockchain-based food traceability systems can address various challenges facing the Chinese market. This is because food and beverage items such as *milk, meat, rice*, and vegetables sold in China are heavily *tainted and contaminated*. For instance, three hundred thousand babies became ill in 2008 after drinking infant formula made from adulterated milk, which led to the death of six infants (Huang, 2014). It is estimated that more than 50% of wines that cost more than $35 per bottle are counterfeit in China and about 30,000 fake bottles of wines are sold every hour in the country (Ambler, 2017).

Blockchain systems have tremendous potential and provide a promising future to strengthen national food safety programs in China. *The* Chinese government has thus made food safety a top priority. Chinese regulators were part of the pilot project run by IBM and Walmart to make the retailer's supply network more transparent by tracing products such as pork and organic food (economist.com, 2017).

Below we briefly describe some of the blockchain-based food traceability programs in China.

a. *Alibaba*

Alibaba's international e-marketplace Tmall uses blockchain and product tagging with a unique QR code assigned to each product. Tmall has partnered with the logistics company Cainiao to use blockchain to track cross-border supply chains.

Food products are tracked and monitored this way and relevant information is made available to consumers. Each step in the supply chain is authenticated and verified. Relevant data such as those related to production, transportation, customs, inspection, and transfer of ownership are secured with blockchain. The blockchain with these proofs is stored by Alibaba. The copies of the records are also stored and validated by other participants (Kaplan, 2019).

Since blockchain solutions are employed with QR codes, it is worth noting that QR codes have advanced which makes counterfeiting impossible *or extremely expensive. Even before introducing blockchain, Alibaba* launched its "Blue Stars" campaign in 2015 for high-end food and other products. The campaign used the next-generation "dotless" QR codes. Participating merchants selling on *Alibaba's online marketplace* Taobao can attach a label containing a QR code with a colorful image to each package to verify the authenticity. A secure scanner developed by software company Visualead is used to scan the QR codes. Each QR code is unique, and cannot be duplicated and brands can customize the code using different combinations of visually appealing images, logos, and different colors (Williams, 2015). Theoretically, it is possible for counterfeiters to sell fake goods with legitimate Blue Stars QR codes. To do so they can buy legitimate products, get enough genuine QR codes and put them on the packages of fakes. However, each item has a unique QR code identifier. When a customer receives the product ordered online and scans the code, it will "burn," which means that each code can be used only one time (Alba, 2018). This means that counterfeiters will have to buy large quantities of legitimate goods to get enough genuine codes. This makes fraudsters' business model less attractive (Erickson, 2015).

Alibaba's blockchain systems have been implemented to track Chinese as well as imported products. In August 2018, Alibaba's online payment affiliate Ant Group signed a strategic cooperation with the Wuchang municipality in China's Heilongjiang province to track the rice supply chain (Sunny, 2018). Alibaba's Tmall and Rookie Logistics are other partners in the project (Kode, 2018). A major goal is to stop counterfeit versions of the Wuchang rice, which is known for its high quality but has limited production and supply. Wuchang is widely counterfeited (Kshetri & Loukoianova, 2019). Each bag of Wuchang rice sold on the Tmall platform displays a QR code with a unique identification number. Consumers can scan this code using a smartphone app before paying for the rice. The details provided include the specific field the rice came from, seeds and fertilizers used to grow the rice, as well as information related to shipments (Chavez-Dreyfuss, 2018).

Alibaba teamed up with New Zealand dairy product maker Fonterra and New Zealand Post to track products imported to China from New Zealand. It also collaborated with Australian healthcare supply firm Blackmores and Australia Post to develop a blockchain-based Food Trust Framework (Hsu, 2018). The goal is to develop a blockchain solution model that participants across food and beverage supply chains can use.

b. *JD.com*

A blockchain system developed by the Chinese e-commerce firm JD.com and inner Mongolia-based food supplier Kerchin connects these two firms. In 2017, the system went live (Xiao, 2017). Kerchin collects and stores data in its supply chain by scanning the barcodes of its products. The information is then entered into the blockchain. After that, any changes in data require a digital signature. Both parties are informed if there is any change or modification in the data (Huang, 2017). JD periodically implements random spot checks at Kerchin's factories to examine the accuracy and validity of information (Huang, 2017).

In early 2018, JD.com announced that it launched a blockchain platform to track imports of beef products to China from Australian meat merchant InterAgri (Zhao, 2018). The initial focus was on high-end beef (Wood, 2018).

c. *Walmart*

In October 2016, IBM, Walmart, and Tsinghua University announced a collaboration to improve the traceability of food products sold to Chinese consumers in order to enhance supply chain transparency and efficiency and ensure food safety. Walmart trial-tested a blockchain-based solution to monitor pork products in the country. The trial took place on a farm operated by a company called Jinluo located in the northeastern city of Lingyi. Jinluo provided data about the pork products such as the farm inspection report and the livestock quarantine certificate (Wong, 2018). These documents were digitized by an industrial personal digital assistant, which is a smartphone-like device in a rugged case. These data were uploaded to the blockchain in real time.

Data provided include the entire history and current location of individual food items such as products, farms, factories, batch number, humidity and storage temperature, and shipping as well as files related to the farm inspection report and the livestock quarantine certificate were secured with blockchain. RFID tags sensors, barcodes, and other sources provided relevant data (Kharif, 2018).

IBM's Food Trust was used, which is based on open-standard and open-governance platform Linux Hyperledger Fabric (Hackett, 2016). This means that the copies of the records are also stored and validated by other peers. It is a SaaS solution that provides immediate access to food supply chain data. Walmart is responsible for setting up its peers to participate in the network. The peers may also include relevant government agencies.

Walmart China commercially launched its Blockchain Traceability Platform in June 2019. By that time, 23 product lines sold in China used the platform, which was expected to increase to 100 by the end of 2019. Walmart aims to have 50% of packaged fresh meats and vegetables and 12.5% of all seafood sales tracked on the platform by the end of 2020 (Duckett, 2019). The company trained about 100,000 employees and suppliers to use the platform. The goal is to make sure that enterprises or consumers can use the system without additional costs (Zhuoqiong, 2019).

In June 2020, *Walmart's* membership-only retail warehouse Sam's Club China launched Sam's Club Blockchain Traceability Platform (creamandpartners.com, 2020). Sam's Club has positioned itself differently in China than in the U.S. It is viewed as a high-end shopping club in China. Sam's Club uses the Chinese Blockchain-as-a-Service (BaaS) platform VeChain's food traceability platform.

In Walmart's trial of a blockchain-based solution to monitor pork products in China, blockchain was enabled to digitally track individual pork products in a few minutes compared to many days taken in the past. Details about the farm, factory, batch number, storage temperature, and shipping can be viewed on the blockchain. These details help assess the authenticity of products. If an item is spoiled or the source of a product is compromised, the system acts proactively. In the case of food contamination, it is possible to pinpoint the products to recall (Yiannas, 2017).

Cryptocurrency, FinTech, and InsurTech

Chinese fintech companies have launched blockchain-based applications to apply in diverse situations and settings such as trade finance, remittances, and reimbursement of employee expenses. For instance, as of mid-2019, Ant Group tracked charity donations of US$50 million using blockchain (Orcutt, 2019).

OneConnect, Ping An Group's fintech company, has developed a blockchain-based platform for small and medium-sized banks (crunchbase.com, 2018). The solution can evaluate potential customers' creditworthiness by extracting a wide range of company data. The company can do so at a low cost and reduce the amount of documentation and manpower needed to process transactions.

Cross-Border Remittance
Efforts have been undertaken by Chinese fintech players to develop low-cost and fast remittance systems based on blockchain. In 2018, Ant Group introduced a blockchain-based cross-border remittance system. AlipayHK and the Philippines-based mobile money service company GCash teamed up to offer real-time money transfer between Hong Kong and the Philippines, which charged significantly lower fees and higher speed and efficiency compared to traditional transfer services (fastcompany.com, 2020). Standard Chartered Bank has been a part of the initiative. Customers make a few clicks with AlipayHK and the money will reach the GCash user's account within seconds. When a user submits a remittance application, all network participants including AlipayHK, GCash, and Standard Chartered Bank get a notification. The sender and receiver can track the money during the entire process (PYMNTS, 2018).

Detecting Fake Receipts in Employee Reimbursement
Detecting fake receipts has been a challenge for Chinese tax collectors, businesses and state-run enterprises. In 2009, Chinese authorities detained 5,134 people and closed 1,045 fake invoice production sites. Likewise, in 2010, Chinese authorities took action against 1,593 criminal gangs and 74,833 enterprises that had filed false invoices. During 2007–2012, the pharmaceutical company GlaxoSmithKline's four senior executives in its China operation submitted fake receipts and defrauded millions of dollars (Barboza, 2013).

Some fintech players aim to address this problem by providing blockchain-based systems for the reimbursement process (known as fapiao in Chinese). In August 2018, Tencent's pilot project tested a blockchain-based feature using WeChat Pay data to inform employers of employees' purchases. Employees can use the system to automatically send transaction data to employers for reimbursement. The feature is expected to bring efficiency in the corporate expense reimbursement process and reduce fraud and tax evasion (Idowu, 2019).

Currently, the reimbursement process requires merchants to issue different receipts for purchases with the employer's taxpayer ID and other information. Merchants manually enter this information to generate receipts and process additional paperwork (Wang, 2018). In December

2018, Tencent announced that qualified merchants can use WeChat's blockchain-backed e-invoices using WeChat Pay to issue invoices (Wang, 2018).

State-Backed Digital Currency
The media have widely reported that China will soon launch a state-backed digital currency known as central bank digital currency (CBDC). As early as 2016, China's central bank, the People's Bank of China (PBOC) had successfully completed a trial of a blockchain-based digital currency prototype system (Fei, 2017). In August 2019, a senior PBoC official said that the country's CBDC was ready to launch. The CBDC pilot was launched in April-August 2020. During that period the People's Bank of China opened 113,300 consumer digital wallets and 8,859 corporate digital wallets in three cities—Shenzhen, Suzhou, and Xiong'an, which processed US$162 million in 3.1 million transactions (Hui, 2020). It explored the benefits of CBDC to the Chinese economy including and the ease with which the Chinese government would be able to trace the financial activities of citizens and businesses.

The CBDC is a two-tier system to replace cash in circulation. The PBoC would issue digital currency to commercial banks. The commercial banks will then issue it to the public (Zhao, 2019). Big Chinese banks (the Industrial and Commercial Bank of China, China Construction Bank, the Bank of China, the Agricultural Bank of China) and fintech companies (Alibaba and Tencent) and Union Pay (an association of Chinese banks) will receive the digital currency to distribute to consumers and businesses (Del Castillo, 2019).

The main reason why the Chinese government wants to do it is because of the ease with which it would be able to trace the financial activities of citizens and businesses (Munster, 2018). The PBOC also hopes that subsequently the cryptocurrency will be made available in foreign countries (Del Castillo, 2019).

The digital currency will handle about 300,000 transactions per second (TPS), which is faster than other cryptocurrencies and card networks. For instance, Visa handles 65,000 TPS (Emem, 2019). According to Facebook's Libra's white paper, the cryptocurrency would handle 1,000 TPS (Zhao, 2019). Likewise, the public Ethereum network's peak throughput 16 TPS (Hanada et al, 2018) and Bitcoin's is 7 TPS (Yli-Huumo et al., 2016). The PBoC official noted that the country's payment networks on the 2018 Singles' Day sale handled 92,771 TPS during the peak time (Zhao, 2019).

The Chinese government is against blockchain networks that follow decentralized and consensus-driven structures. This means that China's state-backed cryptocurrency will be different from well-known decentralized cryptocurrencies such as Bitcoin.

Virtual Currency

In April 2019, the *Cyberspace* Administration of *China (CAC)* released the first list of 197 companies that were approved to conduct business with blockchain (*Global Times*, 2019). There were only two virtual currency projects on the list: VeChain Token (VET) and ParcelX (GPX). Note that virtual currencies are only available in an electronic form and stored and transacted through designated software, mobile, or computer applications, or digital wallets. Transactions of such currencies occur over the Internet.

Of the two approved virtual currencies ParcelX (GPX) was planned to be a crypto-based parcel delivery service, but was not yet launched. The cryptocurrency VET deserves elaboration. It employs so-called native fee delegation feature, which means that the party that benefits from traceability covers the costs of using the network. Consumers and partners that interact with VeChain-powered DApps do not need to hold VET or VTHO. They do not need to pay to write transactions or use the network for other purposes as long as associated gas costs are sponsored, which are specified by the developers (Fenton, 2019). As of July 2020, VeChain Thor was reported to handle 100,000 transactions daily to track supply chains of fashion, food safety, and sustainability (Simmons, 2020).

VeChain utilizes a dual-token economic model. VET is the primary token, which is used to represent value on the network. The utility token VTHO is a "gas" currency, which is required to send transactions or perform actions on the network (https://boxmining.com/vechain/).

VTHO can address issues related to price fluctuations. For instance, if network activity grows and the price of VTHO increases due to higher market demand, it is possible to reduce the cost of a transaction in VTHO terms. In this way, transaction costs in dollar terms can be stable (Simmons, 2020).

InsurTech

Blockchain-based solutions are likely to make insurance-related activities more efficient and fraud-proof. The Chinese healthcare insurance industry suffers from rampant abuses and malpractices that are committed by patients and medical staff. In the Lipanshui city in the Guizhou province, fraud cases were found in 107 of the 135 hospitals and medical centers. All hospitals in Anshun were also found to engage in mismanagement of medical insurance. Some medical staff had provided fake medical records to get payments for treatments, which were not performed (Yan, 2015). Such fraudulent practices can be prevented with blockchain.

The Chinese payment processor and financial services company Ant Group (previously Ant Financial), which is an affiliate company of the *Alibaba* Group, teamed up with the life insurer Taikang Insurance to launch a blockchain-based free health insurance product. When a patient uses Ant Groups's mobile and online payment platform AliPay to pay hospital bills, the patient's identity is linked to Alipay. The electronic invoices issued by the hospital are entered on the blockchain. There is thus no need for AliPay to check their authenticity. After using the prescription once, it cannot be reused. The blockchain-based system thus addresses issues such as false reporting and fraudulent invoices (Ledger Insights, 2018).

4.5   CHAPTER SUMMARY AND CONCLUSION

The Chinese blockchain industry and market offer many interesting features. Blockchain-based solutions are being used to address diverse challenges facing China ranging from trade finance to food safety. For instance, food companies have found that their reputation is likely to increase with these activities. Blockchain-based solutions have facilitated SMEs' access to low-cost working capital.

The quality of entrepreneurial and innovative activities in the Chinese blockchain industry and the market is not clear. Most blockchain businesses have been implemented with low investments. China's recent increase in patent filings in blockchain is often cited as evidence of its innovation prowess. While patents are a good measure of performance, it is too early to see the full effects of these patents and to say whether they lead to Chinese blockchain solutions that dominate the global market.

China's approach to blockchain regulation reflects the tension it faces between using modern technologies to maintain control and using them to stimulate economic growth. The above discussion makes it clear that China has solved this dilemma by effectively designing laws, *regulations, and enforcement mechanisms to* crack down on activities that threaten centralized control. Its government entities have also played key roles in developing blockchain solutions. In the development of the blockchain industry, China has thus broadly followed the same pattern as the Internet. The Chinese government would not allow blockchain implementation without significant modification. In some cases, blockchain applications modified to satisfy China's regulations and policies have lost fundamental elements of the original technology.

REFERENCES

Abadi, J., & K. Brunnermeier, M. (2018). *Blockchain economics* (NBER working paper series). https://www.nber.org/system/files/working_papers/w25407/w25407.pdf

Alba, D. (2018). Alibaba Reveals a New Kind of QR Code to Fight Counterfeits. *Wired*; https://www.wired.com/2015/05/alibaba-reveals-retro-way-fight-counterfeits-qr-codes/.

Allen, Darcy W. E., Berg, C., Davidson, S., Novak, M., & Potts, J. (2019). International policy coordination for blockchain supply chains. *Asia & the Pacific Policy Studies, 6*(3).

Ambler, P. (2017). China is facing an epidemic of counterfeit and contraband wine. *Forbes.* https://www.forbes.com/sites/pamelaambler/2017/07/27/china-is-facing-an-epidemic-of-counterfeit-and-contraband-wine/#5789b7535843

Ayogu, M., & Lewis, Z. (2011). *Conflict minerals: An assessment of the Dodd-Frank Act.* Brookings. https://www.brookings.edu/opinions/conflict-minerals-an-assessment-of-the-dodd-frank-act/

Barboza, D. (2013). Coin of realm in china graft: Phony receipts. *The New York Times.* https://www.nytimes.com/2013/08/04/business/global/coin-of-realm-in-china-graft-phony-receipts.html

Berg, C. (2017). What diplomacy in the ancient near east can tell us about blockchain technology. *Ledger, 2,* 55–64.

Berg, C., Davidson, S., & Potts, J. (2018). *Ledgers.* https://papers.ssrn.com/sol3/papers.cfm?abstract_id=3157421

Blockdata. (2020). Forbes Blockchain 50—Products data deep dive. *Blockdata*; https://blog.blockdata.tech/2020/04/forbes-blockchain-50-products-data-deep-dive/

Brown, C. (2019). *Next big deal for Ethereum competitor VeChain: DNVGL, ASI Group and VeChain team up.* Crypto News Flash. https://www.crypto-news-flash.com/next-big-deal-for-vechain/

Buterin, V. (2018). *Governance, Part 2: Plutocracy is still bad.* https://vitalik.ca/general/2018/03/28/plutocracy.html

Caijing, S. (2018, June 5). The Chinese "gang" manipulating the market—now in EOS? https://carylyne.com/the-chinese-gang-manipulating-the-market-now-in-eos/

Campbell, R. (2016). Babyghost and VeChain: Fashion on the Blockchain. *Bitcoin Magazine.* https://bitcoinmagazine.com/articles/babyghost-and-vechain-fashion-on-the-blockchain-1476807653/

Campbell, R. (2017). Companies adopt Blockchain to enhance security in Asia. *Cryptocoinsnews.* https://www.cryptocoinsnews.com/companies-adopt-blockchain-enhance-security-asia/

Castillo, M. (2019). Blockchain 50: Billion dollar babies. *Forbes.* https://www.forbes.com/sites/michaeldelcastillo/2019/04/16/blockchain-50-billion-dollar-babies/#145541e757cc

Chavez-Dreyfuss, G. (2018). Coca-Cola, U.S. State Dept to use blockchain to combat forced labor. *Reuters.* https://www.reuters.com/article/us-blockchain-coca-cola-labor/coca-cola-u-s-state-dept-to-use-blockchain-to-combat-forced-labor-idUSKCN1GS2PY

Chilnadaily.com.cn. (2020). World's top 10 patent holders by blockchain inventions. *China Daily.* https://www.chinadaily.com.cn/a/202009/18/WS5f63e5c0a31024ad0ba7a3fa_1.html

CoinTelegraph. (2018). Governing decentralization: How on-chain voting protocols operate and vary. *Coin Telegraph.* https://cointelegraph.com/news/governing-decentralization-how-on-chain-voting-protocols-operate-and-vary

creamandpartners.com. (2020). *Walmart china brings together Sam's Club and VeChain to take one step further towards blockchainization with safe food traceability platform.* CREAM. https://creamandpartners.com/walmart-china-brings-together-sams-club-and-vechain-to-take-one-step-further-towards-blockchainization-with-safe-food-traceability-platform/

crunchbase.com. (2018). *OneConnect is a financial technology services company that provides financial technology solutions for small and medium-sized banks.* CrunchBase. https://www.crunchbase.com/organization/oneconnect#section-overview

Del Castillo, M. (2019). Alibaba, Tencent, Five Others to Receive First Chinese Government Cryptocurrency. *Forbes;* https://www.forbes.com/sites/michaeldelcastillo/2019/08/27/alibaba-tencent-five-others-to-recieve-first-chinese-government-cryptocurrency/#3065cb6d1a51

Desai, P., Shabalala, Z., & Daly, T. (2019). Exclusive: London metal exchange to delay ban on tainted metal until 2025—Sources. *Reuters*. https://www.reu ters.com/article/us-lme-metals-sourcing-exclusive/exclusive-london-metal-exchange-to-delay-ban-on-tainted-metal-until-2025-sources-idUSKBN1W 31V2

Duckett, C. (2019). Walmart China turns to blockchain for food safety. *ZD Net*. https://www.zdnet.com/article/walmart-china-turns-to-blockchain-for-food-safety/

economist.com. (2017). If blockchains ran the world: Disrupting the trust business. *Economist*. https://www.economist.com/news/world-if/21724906-trust-business-little-noticed-huge-startups-deploying-blockchain-technology-threaten

Emem, M. (2019). *China's Govt crypto readies huge November launch to steal Libra's thunder*. CCN Markets. https://www.ccn.com/chinas-govt-crypto-readies-huge-november-launch-to-steal-libras-thunder/

Emsley, J. (2018). *China first nation to rank blockchains–Ethereum first, bitcoin thirteenth*. https://cryptoslate.com/china-first-nation-to-rank-blockc hains-ethereum-first-bitcoin-thirteenth/

Erickson, J. (2015). *Alibaba is turning the lowly QR code into a fakes fighter*. Alizila. https://www.alizila.com/alibaba-turning-lowly-qr-code-fakes-fighter-2/

European Union. (2017a). *Regulation (EU) 2017/821 of the European Parliament and of the council*. https://eur-lex.europa.eu/legal-content/EN/TXT/PDF/?uri=OJ:L:2017:130:FULL&from=EN

European Union. (2017b). *Official Journal of the European Union, V. 60*. https://eur-lex.europa.eu/legal-content/EN/TXT/PDF/?uri=OJ:L:2017:130:FULL&from=EN

European Union. (2020). *Study on due diligence requirements through the supply chain*. European Commission Final Report. https://www.business-humanr ights.org/sites/default/files/documents/DS0120017ENN.en_.pdf

Faridi, O. (2020). *Chinese global public blockchain or DLT evaluation index ranks bitcoin in 14th place, EOS ranked at top again*. Crowdfund Insider. https://www.crowdfundinsider.com/2020/04/160112-chinese-global-pub lic-blockchain-or-dlt-evaluation-index-ranks-bitcoin-in-14th-place-eos-ranked-at-top-again/

fastcompany.com. (2020). *Ant financial*. Fast Company & Inc. https://www.fas tcompany.com/company/ant-financial

Fei, A. (2017). *RMB as a digital currency? China successfully completes its blockchain-based digital currency trial*. Lexology. http://www.lexology.com/library/detail.aspx?g=6b306683-2442-4d98-9517-d04da7f812a9.

Feng, C. (2020). Blockchain allowed 17 million people to travel between Guangdong, Macau amid coronavirus pandemic. *South China*

Morning Post. https://www.scmp.com/tech/blockchain/article/3106636/blockchain-allowed-17-million-people-travel-between-guangdong-macau

Fenton, A. (2019). *VECHAIN: Why does good news kill the price of this altcoin?* Micky. https://micky.com.au/vechain-why-does-good-news-kill-the-price-of-this-altcoin/

Fenton, A. (2020). *Blockchain traceability overtakes payments among major corporations.* Cointelegraph. https://cointelegraph.com/news/blockchain-traceability-overtakes-payments-among-major-corporations

Frankenfield, J. (2019). *VeChain definition.* Investopedia. https://www.investopedia.com/terms/v/vechain.asp

Frankenfield, J. (2020). *Consensus mechanism (cryptocurrency).* Investopedia. https://www.investopedia.com/terms/c/consensus-mechanism-cryptocurrency.asp

Frey, C. B., & Osborne, M. (2020). China won't win the race for AI dominance: Authoritarians love data, but innovation matters more. *Foreign Affairs.* https://www.foreignaffairs.com/articles/united-states/2020-06-19/china-wont-win-race-ai-dominance

Garg, P. (2020). *Vechain CEO says China banning crypto trading is actually a good thing.* Cryptoslate. https://cryptoslate.com/vechain-ceo-says-china-banning-crypto-trading-is-actually-a-good-thing/

Gazdecki, A. (2019). *Proof-of-work and proof-of-stake: How blockchain reaches consensus.* https://www.forbes.com/sites/forbestechcouncil/2019/01/28/proof-of-work-and-proof-of-stake-how-blockchain-reaches-consensus/#395f865868c8

Gkritsi, E. (2020). *Alibaba is the top global blockchain patent holder.* Technode. https://technode.com/2020/07/03/alibaba-leads-global-blockchain-patent-but-china-lags-behind-us-and-s-korea/

Global Times (2019). *Cyber administration releases first list of registered blockchain service providers. Golden Times.* http://www.globaltimes.cn/content/1144347.shtml

Hackett, R. (2016). Walmart and IBM are partnering to put Chinese Pork on a Blockchain. *Fortune.* http://fortune.com/2016/10/19/walmart-ibm-blockchain-china-pork

Hanada, Y., Hsiao, L., & Levis, P. (2018). *Smart contracts for machine-to-machine communication: Possibilities and limitations. arXiv* preprint arXiv:1605.01987

Hashimoto, T., & Cho, Y. (2019). China triples US in blockchain patent filings. *Nikkei Asia.* https://asia.nikkei.com/Business/China-tech/China-triples-US-in-blockchain-patent-filings

Heasman, W. (2020). *China dominates global blockchain patent applications.* Decrypt. https://decrypt.co/26589/china-dominates-global-blockchain-patent-applications

Hsu, J. (2018). *Alibaba ups food safety down under via blockchain*. Alizila. https://www.alizila.com/alibaba-ups-food-safety-via-blockchain/

Huang, E. (2017). *Blockchain could fix a key problem in China's food industry: The fear of food made in China*. Quartz. https://qz.com/1031861/blockc hain-could-fix-a-key-problem-in-chinas-food-industry-the-fear-of-food-made-in-china/

Huang, Y. (2014). The 2008 milk scandal revisited. *Forbes*. https://www.forbes.com/sites/yanzhonghuang/2014/07/16/the-2008-milk-scandal-revisited/#fa2aa4e4105b

Hui, A. (2020). *China central bank official reveals results of first digital yuan pilots*. CoinDesk. https://www.coindesk.com/china-central-bank-official-rev eals-results-of-first-digital-yuan-pilots

Hussey, J. (2020). *Expert focus: What supply chain obligations must businesses meet under the EU's conflict minerals Regulation?* Chemical Watch. https://chemicalwatch.com/109471/expert-focus-what-supply-chain-obligations-must-businesses-meet-under-the-eus-conflict-minerals-regulation

Idowu, J. (2019). *Tencent to use blockchain on Wechat for faster refunds of company expenses*. BTCNN. https://www.btcnn.com/tencent-to-use-blockc hain-on-wechat-for-faster-refunds-of-company-expenses/

Jao, N. (2019). *Alibaba, Ping An's OneConnect top blockchain patent rankings: Report*. Technode. https://technode.com/2019/11/21/alibaba-ping-ans-oneconnect-top-blockchain-patent-rankings-report/

Kalathil, S. (2003). China's new media sector: Keeping the state in. *Pacific Review, 16*(4), 489–501.

Kaplan, A. (2019). How Alibaba is championing the application of blockchain technology in China and beyond. *Smartereum*. https://smartereum.com/7630/how-alibaba-is-championing-the-application-of-blockchain-technology-in-china-and-beyond-thu-nov-08/

Kharif, O. (2018). *Wal-mart tackles food safety with trial of blockchain*. Bloomberg. https://www.bloomberg.com/news/articles/2016-11-18/wal-mart-tackles-food-safety-with-test-of-blockchain-technology.

KissPatent (UD). *The current state of blockchain patents*. https://kisspatent.com/blockchain-patents-study

Kode. (2018). Alibaba thinks blockchain will change the world. *Diary Coin*. https://diarycoin.com/alibaba-thinks-blockchain-will-change-the-world/

Kshetri, N., & Loukoianova, E. (2019). Blockchain adoption in supply chain networks in Asia. *IEEE IT Professional, 21*(1), 11–15.

Kshetri, N. (2017). Blockchain's roles in strengthening cybersecurity and protecting privacy. *Telecommunications Policy, 41*(10), 1027–1038.

Kshetri, N. (2019, February 25). *Chinese Internet users turn to the blockchain to fight against government censorship*. Conversation. https://theconversat ion.com/chinese-internet-users-turn-to-the-blockchain-to-fight-against-gov ernment-censorship-111795

Ledger Insights. (2018). *Ant's Alipay uses blockchain to process health insurance claims in seconds*. Ledger Insights. https://www.ledgerinsights.com/blockc hain-alipay-china-health-insurance-claims/

Ledger Insights. (2019). *China dominates global blockchain patents*. Ledger Insights. https://www.ledgerinsights.com/blockchain-patents-global-china-dominates-caict/

Liao, S. (2018). China will soon require blockchain users to register with their government IDs. *The Verge*; https://www.theverge.com/2018/10/22/180 08640/china-blockchain-registration-government-id

Lucien (2020). *300+ Italian companies to track supply chains using vechain*. CryptoTicker. https://cryptoticker.io/en/italy-supply-chain-vechain/

Marinoff, N. (2018). China blocks access to over 120 offshore digital currency exchanges. *Bitcoin Magazine*. https://bitcoinmagazine.com/articles/china-blocks-access-over-120-offshore-digital-currency-exchanges/

McClurg, S., & Talati, K. (2018). *Blockchain & cryptocurrency update: China, trade wars, and NEO*. Crowdfund Insider. https://www.crowdfundinsider.com/2018/09/138785-blockchain-cryptocurrency-update-china-trade-wars-and-neo/

McLaughlin, K. E. (2005). China's model for a censored Internet. *Christian Science Monitor, 97*(210), 1–10.

Munster, B. (2018). *The great blockchain of China*. Derypt. https://decryptme dia.com/3362/the-great-blockchain-of-china

Mwai, C. (2018). Will blockchain fix the mineral traceability woes? *The New Times*. https://www.newtimes.co.rw/news/will-blockchain-mineral-tra ceability

O'Connell, J. (2017). This Chinese firm is already putting businesses on an ethereum-based blockchain. https://www.cryptocoinsnews.com/vechain-is-already-putting-businesses-on-a-blockchain/

Orcutt, M. (2019). China's ubiquitous digital payments processor loves the blockchain. *MIT Technology Review*. https://www.technologyreview.com/s/613478/chinas-ubiquitous-digital-payments-processor-loves-the-blockchain/

Pearson, T. (2019). *Markets are eating the world*. Ribbonfarm. https://www.rib bonfarm.com/2019/02/28/markets-are-eating-the-world/

Peng, T. (2020). *More than 10,000 new blockchain companies established in China in 2020*. Cointelegraph. https://cointelegraph.com/news/more-than-10-000-new-blockchain-companies-established-in-china-in-2020

Peyton, A. (2019). *China bosses blockchain and AI patents*. Fintech Futures. https://www.fintechfutures.com/2019/01/china-bosses-blockchain-and-ai-patents/

POA Network. (2017). *Proof of authority: Consensus model with identity at stake*. Medium. https://medium.com/poa-network/proof-of-authority-con sensus-model-with-identity-at-stake-d5bd15463256

Pwc (2020). *Time for trust: The trillion-dollar reasons to rethink blockchain.* PWC. https://image.uk.info.pwc.com/lib/fe31117075640475701c74/m/2/434c46d2-a889-4fed-a030-c52964c71a64.pdf

PYMNTS. (2018). *AliPay, GCash launch blockchain cross-border remittance service.* PYMNTS. https://www.pymnts.com/news/cross-border-commerce/2018/alipay-gcash-blockchain-cross-border-remittance-philippines/

Reutzel, B. (2017). *Day of demos: Blockchain-IoT consortium kicks off with use cases aplenty.* http://www.coindesk.com/day-of-demos-new-blockchain-iot-consortium-kicks-off-with-use-cases-aplenty/

Simmons, J. (2020). *CREAM co-founder: That's why VeChain is the no. 1 enterprise blockchain.* Crypto News Flash. https://www.crypto-news-flash.com/cream-co-founder-thats-why-vechain-is-the-no-1-enterprise-blockchain/

Sunny, S. (2018). *Alibaba, e-commerce giant considers blockchain for its T-mall with with Cainio.* https://tinyurl.com/ycn5gesl

Taylor, P. (2018). *Alibaba goes live with blockchain food tracking system.* Security Pharma. https://www.securingindustry.com/food-and-beverage/alibaba-goes-live-with-blockchain-food-tracking-system/s104/a7519/?cmd=PrintView&nosurround=true

Terrill, R. (2005). What does China want? *Wilson Quarterly, 29*(4), 50–61.

Thomas, D. (2020). Countdown to EU conflict minerals regulation (7 months). *The National Law Review.* https://www.natlawreview.com/article/countdown-to-eu-conflict-minerals-regulation-7-months

Thornhill, J. (2019). Formulating values for AI is hard when humans do not agree. *Financial Times.* https://www.ft.com/content/6c8854de-ac59-11e9-8030-530adfa879c2

VeChain. (2019). *VeChain is co-founder of The Belt and Road Initiative Blockchain Alliance (BRIBA).* VeChain. https://www.prnewswire.com/news-releases/vechain-is-co-founder-of-the-belt-and-road-initiative-blockchain-alliance-briba-300979679.html

Vechain. (2020). *Toolchain.* Vechain. https://www.vechain.com/product/toolchain

vechaininsider.com. (2019). *Walmart announces VeChain powered food safety platform.* VeChainInsider. https://vechaininsider.com/partnerships/walmart-announces-vechain-powered-food-safety-platform/

VePress. (2019). *The gateway to the VeChainThor Blockchain — Built for everyone.* VePress. https://vepress.org/article/IwJ4xsbQO

Vidrih, M. (2018). *The blockchain international standardization organization: China will dominate.* Good Audience. https://blog.goodaudience.com/the-blockchain-international-standardization-organization-china-will-dominate-b7d423904bfb

Wang, C. (2018). *Tencent integrates blockchain e-invoicing with WeChat*. 8BTC. https://news.8btc.com/tencent-integrates-blockchain-e-invoicing-with-wechat

Williams, D. (2015). *Alibaba is using attractive QR codes so you can check if products are authentic*. TheNextWeb. https://thenextweb.com/apps/2015/05/18/alibaba-is-using-attractive-qr-codes-so-you-can-check-if-products-are-authentic/

Wilmoth, J. 2018. *PBoC digital currency director says blockchains need centralization J*. https://www.ccn.com/pboc-digital-currency-director-says-blockc hains-need-centralization

Wong, J.I. 2018. *Walmart the biggest retailer wants to bring blockchains to the food business*. Quartz. https://classic.qz.com/perfect-company-2/1146289/the-worlds-biggest-retailer-wants-to-bring-blockchains-to-the-food-business/.

Wood, A. (2018). *Chinese retail giant to use blockchain to track beef, prove food safety*. CoinTelegraph. https://cointelegraph.com/news/chinese-retail-giant-to-use-blockchain-to-track-beef-prove-food-safety

Xiao, E. (2017). *Alibaba, JD Tackle China's fake goods problem with blockchain*. Tech in Asia. www.techinasia.com/alibaba-jd-ecommerce-giants-fight-fake-goodsblockchain.

Yan, T. (2015). *High levels of medical insurance fraud found in rural areas of southern Chinese province*. http://www.scmp.com/news/china/policies-pol itics/article/1850183/high-levels-medical-insurance-fraud-found-rural-areas

Yang, D. L. (2001). The great net of China. *Harvard International Review, 22*(4), 64–69.

Yiannas, F. (2017). *A new era of food transparency with Wal-Mart center in China*. Food Safety News. http://www.foodsafetynews.com/2017/03/a-new-era-of-food-transparency-with-wal-mart-center-in-china/#.WOB65m cVjIU

Yli-Huumo, J., Ko, D., Choi, S., Park, S., & Smolander, K. (2016). Where is current research on blockchain technology?—A systematic review. *PLOS*. https://journals.plos.org/plosone/article?id=10.1371/journal.pone. 0163477

Zhao, W. (2018). *JD.com to track beef imports using blockchain platform*. Coindesk. https://www.coindesk.com/jd-com-to-track-beef-imports-using-blockc hain-platform

Zhao, W. (2019). *China's digital fiat wants to compete with bitcoin—But it's not a crypto*. Coindesk. https://www.coindesk.com/is-chinas-digital-fiat-a-crypto currency-heres-what-we-know

Zhuoqiong, W. (2019). Walmart China launches blockchain platform to help shoppers track products. *China Daily*. http://www.chinadaily.com.cn/a/201 906/26/WS5d130a01a3103dbf1432a5e3.html

CHAPTER 5

Blockchain in the Mineral and Metal Industry

Abstract This chapter examines blockchain's roles in promoting ethical sourcing in the mineral and metal industry. It gives detailed descriptions of how blockchain-based supply chain networks' higher density of information flow and a high degree of authenticity of information can increase supply chain participants' compliance with sustainability standards. It gives special consideration to blockchain systems' roles in overcoming the deficits in the second-party and third-party trust. It also demonstrates how blockchain-based supply chain networks include outside actors and configures the supply chain networks in a way that enhances the empowerment of marginalized groups.

Keywords Cobalt · Corporate social responsibility · Dodd-Frank Wall Street Reform and Consumer Protection Act · Environmental · Social and governance · Lithium ion batteries

5.1 Introduction

A number of serious environmental, social, and governance (ESG) issues have been identified in the mineral and metal industry supply chains (Lèbre et al., 2020). Mineral extraction activities' adverse impacts include environmental degradation, natural resources exploration, exploitation

© The Author(s), under exclusive license to Springer Nature Switzerland AG 2023
N. Kshetri, *Blockchain in the Global South*,
https://doi.org/10.1007/978-3-031-33944-8_5

of child workers, human rights violations, population displacement, and violent conflicts (UNSDSN, 2016). Most serious ESG violations often occur deeper down in supply chains (Sedex, 2013) affecting marginalized groups such as artisanal and small-scale miners (ASMs).

The *wide media coverage has* brought to light highly unethical practices in mineral and metal extraction activities including the use of child labor, human rights violation and environmental damages. This is especially true for cobalt. About 10–20% of lithium-ion batteries consist of cobalt (Nelson, 2019). A battery of an electric car requires 10–20 kilograms of cobalt (Wolfson, 2019). About two-thirds of cobalt used in the world is mined from the DRC. There are reports that children as young as six work in the mines with unsafe working conditions (Taylor, 2020). The wages are as low as $0.75/day. A large number of preventable deaths have been reported in the country's cobalt mining industry. Due to such issues, cobalt is also referred to as "blood cobalt" (LeVine, 2018) and is also known as the "Blood Diamond of Batteries"(Airhart, 2018).

As noted in Chapter 4, Western countries have started responding to some of the serious ESG issues. For instance, countries that are covered under the Dodd–Frank Wall Street Reform and Consumer Protection Act include South Sudan, Uganda, Rwanda, Burundi, Tanzania, Malawi, Zambia, Angola, Congo, Central African Republic, and the DRC (Ayogu & Lewis, 2011). Many jurisdictions in Europe and elsewhere have similar legislation. Despite these regulatory developments, addressing ESG issues in the mineral and metal industry is not an easy task. The mineral and metal industry supply chains have unique, unusual, and idiosyncratic characteristics. For instance, mineral and metals often change physical characteristics, chemical composition, and other features along the supply chain. Commenting on the challenges involved in tracking tantalum, which is used for making capacitors for devices such as smartphones and laptops.

Blockchain *has the potential to address many of the ESG issues in supply chains. Unsurprisingly* some entities of governance such as the European Union (EU) have recommended that the use of technology such as blockchain be explored to enhance supply chain visibility in these countries (European Union, 2020). *Emphasizing the seriousness and* urgency of *ESG* issues in this industry and blockchain's potential, Everledger CEO Leanne Kemp said: "We saw that the next most potentially conflicted supply chain is going to be in rare earths and batteries. We're not interested in tracking lettuce. That's not where the world needs us to be" (Allison, 2020).

Mineral and metal industries in less represented African economies such as Rwanda, the DRC, and Sierra Leone have been picked up with enthusiasm by commercial organizations developing blockchain solutions to help multinational corporations (MNCs) to track their CSR activities and performance. These MNCs' adoption of blockchain to track corporate social responsibility (CSR) are shaped by diverse motivations and circumstances.

Different materials in supply chains pose different levels of difficulties in implementing blockchain to ensure tamper-proof tracking of production and distribution processes. For instance, it is relatively simple to track diamonds compared with ores such as cobalt and columbite-*tantalum* (coltan). Each individual diamond cut has unique elements, which can be translated into data attributes to ensure the immutability of every transaction (Forbes Africa, 2018). On the other hand, coltan needs to be refined to produce tantalum. *Minerals that rely on smelting and refining process* technology are more difficult to track. The refining process increases the risk of clean batches of materials being mixed with other batches of products potentially *containing conflict minerals* (Uwiringiyimana & Lewis, 2018). The complexity also increases as the number of players in a supply chain increases.

Among the minerals, cobalt has attracted the most regulatory and media scrutiny. n a global basis, the leading use of cobalt is in rechargeable battery electrodes (usgs.gov, 2020). We thus looked at the status of blockchain adoption among the key players in the lithium-ion *battery industry and the automobile industry.* In the lithium-ion battery industry, the two biggest companies in 2019 were China's Contemporary Amperex Technology Co. Limited (CATL) and South Korea's LG Chem (Bohlsen, 2019). LG Chem is a member of the Responsible Sourcing Blockchain Network (RSBN), which is an industry collaboration that aims to support sustainable and responsible sourcing and production practices. The traceability-as-a-service (TaaS) provider Circulor operates a blockchain platform across *CATL*'s supply chain (Rolander, 2019). Volvo *is a participant in CATL*'s supply chain that Circulor operates. In the future, Volvo plans to apply RSBN to other minerals found in batteries such as nickel and lithium (rcsglobal.com, 2019).

5.2   Key Issues in Supply Chains

Supply Chain Governance and Current Challenges

The term governance can be defined as a process by which an organization or a society steers, coordinates, and manages itself (Norris, 2000). There are several strategies and governance mechanisms that firms can use to manage supply chain relationships and governance. *Two categories of mechanisms* are considered (Heide, 1994). First, a firm can select exchange partners that have the ability as well as a willingness to support its strategy. For example, a company can require a potential contractor's participation in its formal qualification program. Second, incentive structures can be designed to reward desirable behaviors and penalize noncompliance (Williamson, 1983).

These are, however, easier said than done. In this section, we discuss various challenges that confront supply chains.

Information, Communications, and Knowledge Gaps

Various gaps hinder sustainable and responsible supply chain governance. The needs related to reliable, comprehensive, authentic, and credible information about sustainability impacts at various phases of supply chains are not fulfilled. These gaps are referred to as *information and knowledge gaps* (Boström et al., 2015). In supply chain relationships, it is a common practice for buyers to evaluate the operational performance of suppliers. The process is referred to as vendor rating (Luzzini et al., 2014). Information required to perform vendor ratings is unavailable for small vendors from developing countries.

Implementation Challenges

Standards that have been formulated and various sustainability principles and criteria that have been defined are not necessarily followed by supply chain participants. Ensuring supply chain participants' compliance with strict principles and guidelines is often a challenging task (Boström, 2015). These lead to *compliance or implementation gaps*. To take an example, the global apparel retailer C&A requires its suppliers to respect its ethical standards which include fair and honest dealings with employees, sub-contractors, and other stakeholders (Graafland, 2002). There are, however, implementation challenges due to the technical impracticality of assessing various stakeholders' sustainability practices (Kshetri, 2020).

Power and Accountability Issues

An unequal distribution of power or the lack of power symmetry among various supply chain actors hinders the development of responsible and sustainable governance in supply chains. Such gaps are known as power gaps. In an institutionalized relationship, when decision-makers exercise power and control over certain groups but are unwilling to fulfill their responsibility, a problem of "accountability deficits" arises in a governance system (Macdonald, 2007). When big firms use coercive power, less powerful actors such as small firms and workers may be left vulnerable (Forster & Regan, 2001). For instance, some workers are forced to work in low-wage informal sectors.

Misleading Persuasive Practices

Many firms engage in unsustainable activities under the name of sustainability (Blühdorn, 2007). Various standards follow narrow definitions of sustainability, which tend to favor powerful stakeholders and fail to reflect the concerns of marginal groups (Bush et al., 2015).

For instance, mining industry is responsible for over a quarter of global carbon emissions, and has displaced communities that are vulnerable to climate change (London Mining Network, 2019). However, this industry cites projected critical metals demand and frames itself as an actor to fulfill the demand to justify new projects and attract investment and framing (War on Want & London Mining Network, 2019). There are thus *credibility* or *legitimacy gaps*. Such gaps can be attributed to the creation of an illusion of improvement without being accompanied by real improvement and low degree of transparency (Mol, 2015).

Lack of Connection with Actors Outside Supply Chains

Current supply chains fail to engage a wider set of actors and institutions. Initiatives such as the Fair Trade systems have been designed as standalone, *discrete systems of* supply chain governance *that lack* the wider components to connect actors outside supply chain institutions. Due to *the lack of integration with actors outside a supply chain,* such supply chain governance systems thus have limited ability to improve the well-being of workers and producers (Macdonald, 2007). There is *the lack or ineffectiveness of watchdog* organizations to make sure that companies follow sustainability standards.

Such problems become even more complicated when unaccountable power groups make decisions affecting marginalized groups. In such cases, a decision-making unit has only partial control, and thus only partial responsibility for outcomes. They also lack the ability to influence external decision makers to fulfill their responsibility (Macdonald, 2007).

Responsible Supply Chain Governance
to Empower Marginalized Workers

Three necessary conditions have been identified in the context of the global coffee industry for ethical consumption campaigns (Macdonald, 2007). *First,* relevant decision makers in the global North must accept their expanded responsibility to *address the disempowerment problem* that has affected these disadvantaged groups. Second, institutional capabilities must be strengthened so that the decision makers do everything that needs to be done in order to empower the marginalized workers and producers. Third, the marginalized groups must be given a voice and represented in decision-making processes so that they can exercise some control over the institutional transformation processes.

The above observation underscores the importance of changes at two levels to fight the disempowerment and marginalization of disadvantaged groups in this industry. First, interactions and relationships within the supply chains themselves must be changed. Second, some mechanisms must exist to ensure the participation of outside actors.

Regarding the first condition, the choice of a governance models in a supply chain is a function of how various participants are connected to each other and their relative power and position within the network (Rowley, 1997). Two key supply chain characteristics deserve mention: centrality and density. Centrality refers to an actor's position relative to others. A high degree of centrality leads to a more prominent intermediary position (Rowley, 1997). Supply chain density provides a measure of actors' interconnection along a value chain. In a supply chain network with a high number of connections among nodes, an actor will receive more attention from other participants. The increased sharing of information among participants would lead to increased monitoring of organizations (Neville & Menguc, 2006). Organizations face pressures to comply with stakeholder expectations.

As to the second condition, a coordinated set of actions from a range of actors beyond as well as within supply chain institutions can improve the well-being of marginalized workers and producers (Macdonald, 2007).

Theoretically, the problem of disempowerment can be addressed by reconfiguring responsibilities across a plurality of decision makers within and beyond supply chains. That is, partial and shared responsibilities should be appropriately allocated and coordinated. However, the initiatives taken so far have largely failed to develop transparent means to define the boundaries of partial responsibilities. There has also been a lack of institutional mechanisms to strengthen the coordination among relevant actors (Macdonald, 2007).

There have been some encouraging developments with regard to the roles of actors beyond supply chains in influencing ESG issues in supply chains. In the past few decades, various social groups have responded to what they view as a "distorted and unjust" governance system in global production and trading systems (Macdonald, 2007). Global supply chains thus have been subjected to intense criticism and recent works of social groups have focused on the harshness of working conditions.

Blockchain's Roles in Sustainable Supply Chain

An emerging application of blockchain has been in demonstrating sustainability. More broadly, blockchain can help achieve various supply chain goals including sustainability (Kshetri, 2018). Blockchain can make it possible to make sustainability-related indicators transparent, quantifiable and more meaningful (Kshetri, 2018).

Blockchain's features such as decentralization and immutability make it an ideal tool to improve supply chain traceability and address various shortcomings of traditional supply chains (Kim & Laskowski, 2018). Immutable data related to nature, quality, quantity, location and ownership, and other characteristics can address such issues. While non-blockchain supply chain information systems can uniquely identify products they perform poorly in traceability. Traditional supply chain information systems *suffer from data silos*—some supply chain *data* are accessible by some participants but are *isolated* from others. In order to trace ingredients across supply chains, data must be accessible to relevant parties.

Improving the governance structures in supply chains is another key mechanism by which blockchain affect sustainability. This technology can provide *visibility and provenance* f and facilitate the automation of tasks such as payments, and settlements (Narayanaswami et al., 2019). For instance, *blockchain can be used to create a* supply chain *map showing the flow of transactions and information, which* can help identify the weakest links and understand the degree and nature of risks and threats involved (Min, 2019). All these can lead to a reduction in opportunistic behavior.

Blockchain can *make some* intermediary *tasks redundant*. Such disintermediation can transform supply chains by making it possible to conduct transactions without relying on a third party. Distributed trust based on the consensus of a network of participants can replace trust produced by third parties (Francisco & Swanson, 2018), which can help reduce transaction costs and facilitate market-oriented practices (Cole et al., 2019).

Blockchain can help increase the authenticity of product information provided to consumers, which can increase their confidence in the product. Blockchain thus facilitates product traceability, which increases supply chain transparency and enhances consumers' perception of a firm's sustainability practices. Regarding the mechanisms, ultimatum game experiments have shown that in order to punish unfair practices, individuals are willing to give up some monetary benefits (Camerer & Thaler, 1995). When there is the possibility that unfair behaviors lead to punishment, firms are less likely to engage in such behaviors. A practical challenge, however, is that it is often difficult to assess the fairness of some participants' behaviors. Blockchain-based transparency can expose unfair practices.

5.3   Some Key Players Involved in Blockchain Deployment in the Mineral and Metal Industry

Table 5.1 provides a brief description of some key players involved in blockchain deployment in the mineral and metal industry.

De Beers

As explained in Chapter 1, De Beers' blockchain solutions track diamonds as they move from the mine to cutter and polisher, and then to jewelers. The blockchain platform Tracr was launched in 2018 to establish provenance, authenticity, and traceability in diamond supply chains (Sabine, 2019). The platform was developed with Boston Consulting Group's Digital Ventures using the Ethereum blockchain. As of 2020, more than 30 participants including Signet Jewelers, Chow Tai Fook Jewelry Group and diamond mining company Alrosa were using Tracr (Sabine, 2019).

Table 5.1 A brief description of key players involved in blockchain deployment in the mineral and metal industry

Organization	Blockchain used/key users	Sample blockchain projects/performance indicators
RSBN	Hyperledger Fabric overseen by RCS Global	2019: initial test: 1.5 ton batch of cobalt was tracked from Huayou's mine site in the DRC to refinement facility in China, then to LG Chem's cathode and battery plant in South Korea, and finally to a U.S.-based Ford plant
Circulor	Oracle's blockchain platform (OBP)	November 2019: Volvo, CATL and other participants recorded about 28 million material scans and other production events per month
De Beers	Tracr was developed with Boston Consulting Group's Digital Ventures using Ethereum	2020: more than 30 participants, tens of thousands of precious stones are registered per month on Tracr
Everledger	IBM's TrustChain platform built on Hyperledger Fabric	April 2019: recorded the origins of about 2.2 million diamonds

Tens of thousands of precious stones are registered every month on the platform (Debter et al., 2020).

Tracr started with tracking bigger diamonds. The initial test was conducted with 100 large diamonds. In mid-2018, the platform was used to track a rough stone of 2 carats and above (Bates, 2018). In the early test, most of its supply chain activities were owned or controlled by De Beers, which made compliance relatively easy to achieve.

Tracr aims to develop a "Global Diamond ID." Diamonds undergo 3D scans when they are mined, cut, polished, and sold (Debter et al., 2020). *Tracr* assigns each diamond a unique ID. It uses scientific data involving 200 different characteristics such as weight, color, clarity, and photos to uniquely identify each diamond (Bates, 2018).

As discussed in Chapter 1, De Beers launched a GemFair program for ASMs. ASMs are required to identify and manage key risks defined in "OECD Due Diligence Guidance for Responsible Supply Chains from Conflict and High-Risk Areas" in order to participate in the GemFair program (Gemfair, 2019). Among the major requirements is that ASMs need to identify the worst forms of child labor and address them. Compliance with the requirements is ensured through first-party (e.g., a member completes a self-assessment workbook provided by GemFair), second-party (GemFair's monitoring of mine sites bi-annually) and third-party (commissioning a third-party assessment of a sample of mine sites twice a year) verifications.

Tracr identified three major challenges that must be addressed for tracking diamonds in supply chains: (a) determining the features to uniquely identify a piece of rough diamond; (b) determining the features to uniquely identify a piece of polished diamond; and (c) matching a polished piece with the rough piece it comes from. The last step is arguably the most challenging one. It was reported that Tracr successfully tracked 200 different diamonds in its pilot phase (Bates, 2018). The company claims that it uses state-of-the-art AI tools to observe a diamond to determine its supply chain journey (https://www.tracr.com/).

Everledger

The London-based technology enterprise Everledger's blockchain-based solutions are used to *verify provenance of products*. It was first used for tracking rough-cut diamonds.

Everledger uses IBM's TrustChain platform built on the Hyperledger Fabric (Jamasmie, 2019), which is a modular blockchain system that allows organizations to develop products, solutions, and applications based on blockchain. Key components such as consensus and membership services work on a *plug-and-play basis*. It thus allows organizations to conduct confidential transactions without the need for a central authority.

Everledger gives a unique cryptographic ID to each piece of diamond. It does so by storing diamonds' unique identities that are derived from more than 40 attributes. They include the so-called 4Cs—carat, cut, clarity, and color—as well as information about provenance and price. The cognitive analytics systems utilize AI to cross-check data related to regulations, relevant records, supply chains, and IoT to ensure that the gems from conflict regions do not enter the global supply chain (Thibodeaux, 2018). All permissioned parties in the supply chain have access to data.

As noted above, one of the necessary conditions for empowering marginalized groups would be to give them a voice and represent them in decision-making processes (Macdonald, 2007). *In this regard*, Everledger and Swiss-based jewelry retailer Gübelin provide a no-cost solution to track colored gems produced or manufactured by ASMs. ASMs can use Everledger's blockchain platform to create traceability and document retention for free (Cholteeva, 2019). Being part of the blockchain network would allow them to exercise some control over the institutional transformation processes (Macdonald, 2007).

Everledger launched its Diamond Time-Lapse Protocol in 2017, which provides real-time data related to origin, cutting and polishing, artisans' work, and certification. The protocol has two user interfaces: Manufacturer and Retailer User Interface and Consumer User Interface. The Manufacturer and Retailer User Interface allows manufacturers to capture data as a piece of diamond moves through the manufacturing processes. Retailers can record relevant retail information when the diamond reaches the point of sale. The Consumer User Interface is a mobile application for iOS and Android operating systems. Customers can log in to the system to view the complete provenance report of their purchased diamonds (IDEX, 2018).

The Responsible Sourcing Blockchain Network (RSBN)

The RSBN is an industry collaboration that aims to support sustainable and responsible sourcing and production practices. RSBN members include IBM, Ford, Volkswagen, Huayou Cobalt, Fiat Chrysler, Volvo, LG Chem, and British-Swiss commodities trading company Glencore. The RSBN blockchain platform is built on Hyperledger Fabric and is overseen by responsible sourcing group RCS Global (rcsglobal.com, 2019).

The project has been implemented in southern Congo (Ross & Lewis, 2019). At the point, where cobalt is bagged and tagged, the miner (Huayou) adds data into the blockchain. For successive stages and key events such as smelting and refining, data related to inputs and corresponding outputs are added to the blockchain. New pieces of information related to shipping and other details are added from partners along the supply chain route. The record is automatically updated each time a transaction is added and made visible to the permissioned participants in real time (Devanesan, 2020). The idea is to allow regulators as well as end users to verify the data (Baydakova, 2019).

In an initial test, the RSBN demonstrated the use of blockchain to track cobalt produced at Huayou's mine site in the DRC. The flow of 1.5-ton batch of cobalt in the supply chain was traced through mines in the DRC which was refined in China (Khatri, 2019). The refined mineral was then sent to LG Chem's cathode and battery plant in South Korea, and then to a U.S.-based Ford plant (MENA Report, 2019). The three-continent journey of the cobalt refinement process took over five months (Nelson, 2019). In the future, RSBN users aim to expand to other metals. For instance, Volvo Cars plans to apply RSBN to other minerals found in batteries such as nickel and lithium (rcsglobal.com, 2019).

Circulor

Circulor utilizes the Oracle's blockchain platform (OBP). The OBP is based on the Linux foundation's Hyperledger Fabric as the foundational technology. Oracle is a BaaS provider. OBP sets up, manages, and maintains the blockchain platform for enterprises (Acharya, 2019).

OBP is combined with AI algorithms to perform due diligence, and identify data anomalies and actions that need further investigation. Data captured include the ore's origin, attributes (e.g., weight and size), the chain of custody, and information to establish supply chain participants' actions comply with globally recognized supply chain guidelines (Wolfson, 2019).

The application's *field test was carried out* for Tantalum mined in Rwanda and then for Cobalt used in Volvo Cars' electric vehicle batteries (Hall, 2019). For the project in Rwanda, Circulor teamed up with the government of Rwanda and Power Resources Group (PRG), which has mining and refining operations in Rwanda and Macedonia (Côme Mugisha, 2019). In 2014, Rwanda accounted for 50% of the production of global tantalum concentrates (Sanderson, 2015). As of 2019, Volvo, CATL, and other participants recorded about 28 million material scans and other production events per month on the Oracle platform (Wolfson, 2019).

The combination of AI and blockchain can be an effective way to address *information and knowledge gaps, which* represent a major challenge that supply chains are facing (Boström et al., 2015). There is a lack of reliable, authentic, and credible information about sustainability impacts at various phases of supply chains. Reliability and authenticity of *data in the first mile* of the supply chain, which is the most crucial step in assuring the quality of the ore (Brugger, 2019), are a key challenges. For instance, blockchain systems can be corrupted if the government agents whose role is to tag bags collude with smugglers and enter incorrect data

(Cant, 2019). In Circulor's system, miners enter the data, whose identities are confirmed with facial recognition software (Ross & Lewis, 2019).

5.4 Blockchain's Role
in Promoting Ethical Sourcing

The cases discussed above illustrate the roles of blockchain systems in *identifying, tracing, and tracking* relevant information in mineral and metal supply chains. The systems ensure that data are shared in a tamper-proof way and are accessible to relevant parties, which makes it possible to trace products across multiple tiers of a supply chain.

Blockchain systems to trace metals such as diamonds are simpler to use compared to minerals such as cobalt. For instance, Everledger stores diamonds' unique identities derived from more than 40 attributes. Minerals such as cobalt traced by RSBN and Circulor go through complex stages such as smelting and refining, which makes it difficult to adopt a foolproof procedure. Everledger also has a longer experience in providing traceability solutions. Everledger was established in 2015 and started tracing the provenance of diamonds using a permissioned blockchain the same year. Circulor was founded in 2017.

In each of the cases analyzed above, major technologies other than blockchain are an integral part of traceability systems. They include OBP's AI algorithms to perform due diligence in Circulor's system, IBM's exploration of chemical analysis using AI to pinpoint the origin of cobalt in RSBN, AI tools to observe a diamond to determine its supply chain journey in De Beers' system and Everledger's use of AI to cross-check data related to regulations, relevant records, supply chain, and IoT to ensure that the gems from conflict regions do not enter the supply chain (Thibodeaux, 2018). In Table 5.2, we present some of the roles of blockchain and other major technologies in enhancing traceability.

A Decentralized Network with a High Supply Chain Density

A supply chain system can be viewed as trusted relationships that are maintained with various ledgers (Berg et al., 2018). In the non-blockchain world, these ledgers and records are *maintained by centralized* entities. As an institutional governance mechanism, blockchain turns centralized management of records upside down by enabling decentralized governance of such records (Allen et al., 2019). In De Beers' system,

Table 5.2 The uses of blockchain and other major technologies in identifying, tracing and tracking relevant information

Identifying, tracing and tracking *information about*	*Technology*	
	Blockchain	*Other major technologies*
People and organizations	Cryptography-based digital signatures verify identities of participants Provides a foolproof method of verifying certain sustainability indicators such as payments made to miners' wallets	Circulor: Facial recognition software: confirm the identities of miners; machine learning and aerial Imagery: determine whether a mining company has employed children Satellite data: verifies that a mine is working AI: checks if supply chain participants' actions comply with globally recognized supply chain guidelines
Minerals and metals	Data are stored and shared in a tamper-proof way and are accessible to all relevant parties (e.g., Everledger stores diamonds' unique identities that are derived from more than 40 attributes)	IBM's planned AI solution: chemical analysis to pinpoint the origin of cobalt Tracr: AI tools to observe a diamond to determine its supply chain journey De Beers: GPS locations for each diamond found Circulor: smartphones with GPS capability to pinpoint the location where the ore was tagged

for instance, all relevant participants such as miners, cutters, polishers, the validator, and jewelers receive information about all transactions. In this way, blockchain-based supply chain models are characterized by low centrality and high supply chain density. Due to a high degree of information flow and a low relative dominance of a given actor, it will be in the interest of supply chain participants to comply with sustainability standards. The tendency of manufacturers and retailers to exploit the information asymmetry to increase profits by providing false information about their products can be addressed with the deployment of blockchain.

Impacts on Trust and Enforcement

There are three ways to produce trust: (1) institution-based trust is linked to institutions such as government bureaucracies and other formal mechanisms, trade associations and professions; (2) process-based trust is produced from the engagement in trustworthy relationships; and (3) characteristic-based trust is generated by identifiable attributes that are linked with trustworthy behavior (Zucker, 1986).

Institutional trust-producing structures are not well-developed in many resource-rich African countries. For instance, due to factors such as corruption and political patronage, there is a low degree of trust in the DRC's government and its institutions. In general, third-party enforcement mechanisms, which are often formal coercive enforcement measures by the state, have been relatively ineffective in the developing world. This increases the relative importance of other types of trust and enforcement.

Regarding process-based trust, which is related to the second-party trust, there have been instances of untrustworthy transactions. Most companies rely on a paper-based certification. UN experts have documented cases in which tags used to identify clean minerals were stolen in eastern Congo and sold to smugglers. Ore from blacklisted mines was sold as responsibly sourced (Ross & Lewis, 2019). The artisanal extraction of cobalt in the DRC has also been linked to toxic harm to vulnerable local communities (Nkulu et al., 2018). There is thus the lack of process-based trust due to the lack of some actors' engagement in trustworthy relationships.

The above problem is the result of the lack of the second-party trust. The questions here include who the party is and whether they behave in a way that is mutually agreed upon.

The lack of institutional trust and process-based trust means that firms in the African metal and mineral industry are essentially left with only characteristic-based trust. That is, blockchain-based supply chains to track minerals are viewed as having attributes that are linked with trustworthy behavior. The challenges related to the deficits and the second and third-party trust can potentially be addressed by incorporating blockchain in supply chains.

The problem of trust deficit can thus be dealt with using a methodology *based on natural science. For instance, t*he position information as to the location of mineral extraction sites can be determined from various sources. De Beers' program records GPS locations for each diamond found. Likewise, in Circulor's system to monitor mines in Rwanda, *smartphones with GPS* capability are used to pinpoint exactly the location where the ore was tagged (Ross & Lewis, 2019).

Another important consideration is that blockchain systems represent a trade-off between efficiency and trust. Private, permissioned blockchains remove the need for slow *and* cumbersome verification process that completely decentralized blockchains such as bitcoin use. Private blockchains such as Hyperledger Fabric used by RSBN, Circulor, and Everledger are thus much faster and more efficient than the public, permissionless systems and thus are better suited in the context of supply chain transactions that require handling large volumes of data in real time (Burns, 2016). There is, however, a risk that a large player can create a monopoly on the global mineral supply chain tracking initiatives using private, permissioned blockchains (Gleeson, 2019). Public blockchains could provide a safeguard against such risks. For this reason, public blockchain would be viewed as more justifiable than private permissioned blockchains if the trust issue is extremely critical.

In a blockchain platform initiated by an individual company such as De Beers [2], which is also used by its competitor Alrosa, a few additional considerations need to be addressed. In a situation such as this, there is, what is referred to as an "asymmetrical data problem" (Thompson, 2020). The idea is that the company which owns the platform is perceived to derive more value from data exchanges in the platform, especially if the platform is based on a private, permissioned blockchain (Thompson, 2020). Compared to solutions provided by a third party technology company or developed by a consortium, private blockchain platform's trust-producing roles are likely to be more problematic when such platforms are developed and controlled by a private company.

A key mechanism to manage supply chain relationships and governance *is to* design incentive structures to reward desirable behaviors and penalize noncompliance. A challenge is to assess (non-)compliance. For instance, when minerals are smelted, they are often combined with metals from various sources. This increases the difficulty of tracking. Companies are looking at advanced technological solutions such as AI to prevent such practices. It was reported that IBM was exploring the possibility of performing chemical analysis using AI to pinpoint the origin of cobalt. The goal is to ensure that "clean" batches of cobalt are not smelted with minerals that have been sourced unethically (Lewis, 2019).

Such systems have been or are being undertaken to track p*eople and organizations as well*. For instance, when a registered mining company that has a concession applies to use Circulor's mine-to-manufacturer traceability of Tantalum, the coordinates of the mine's operations and its historical production are entered into the system. Satellite data is used to verify that the mine is working (Burbidge, 2019). Circulor's plan is to use machine learning models and aerial imagery to ensure that child labor has not been used in the production process (Kapilkov, 2020a).

Supply Chain Networks with Reconfiguration of Responsibilities

A drawback of the existing governance arrangements in modern supply chains is their standalone and *discrete nature with a low degree of integration with actors outside the supply chain* (Macdonald, 2007). *Blockchain-based system* is superior in the sense that outside actors are connected in the supply chain systems. For instance, TrustChain, which is used by Everledger has included Underwriter Labs (UL), as an independent third-party verifier. The idea is to increase the confidence of the TrustChain platform (Hill, 2018). The fact that information in the ledger is verified by third-party verifiers such as UL further strengthens the authenticity of information. Likewise, Everledger's Diamond Time-Lapse Protocol allows customers to view the complete provenance report of their purchased diamonds (IDEX, 2018). In the same vein, RCS Global is the validator of the RSBN (Todd, 2019). Also, the idea behind RSBN is also to allow regulators and end users to verify the data (Baydakova, 2019). government officials put barcoded tags on the sacks of tantalum ore (Ross & Lewis, 2019). The addition of such nodes leads to a further increase in supply chain density and supply chain participants' propensity to comply with sustainability standards.

Blockchain can fulfill the necessary conditions for sustainable supply chain management that can empower marginalized groups. For instance, one way to improve accountability mechanisms would be to disaggregate responsibilities between relevant decision makers and coordinate decision-making processes to achieve a given goal (workers' and producers' well-being) (Macdonald, 2007). The required coordination among the different actors within and beyond the supply chain institutions can be achieved by combining blockchain and other technologies.

One way to address the problem of disempowerment is to reconfigure the allocation of responsibility (Macdonald, 2007). Due to corruption and poor enforcement of the rule of law, *the governments of mineral and metal-originating countries such as the DRC have not been able to deal with* human rights and child labor problems. These governments do not have the same incentives and pressures to be accountable as Western MNCs.

From the perspective of Western MNCs, one way to deal with enforcement challenges and the third-party trust deficit would be to take enforcement responsibility themselves. In July 2020, Volvo Cars' venture capital investment arm Volvo Cars Tech Fund announced an investment in Circulor (Volvo Cars, 2020). Three other investors—SYSTEMIQ, Total Carbon Neutrality Ventures and Plug & Play—joined Volvo Cars Tech Fund in that round (Kapilkov, 2020a). Circulor plans to use funds from new funding sources to train and improve its machine learning models so that they can distinguish between children and adults with a high level of accuracy. With such a capability, the firm hopes to be able to use aerial imagery to determine whether a mining company has employed children in its operations (Kapilkov, 2020b). In this way, the enforcement responsibility shifts from the local government to the blockchain system. Blockchain and other major technologies can help redesign the responsibilities of various actors so that it will be possible to more effectively address the challenges of disempowerment.

Measures for Giving Voice to Marginalized Groups

In the governance systems of traditional supply chains, weak and peripheral actors' lack of voice and resources is among the main reasons that would prevent them to comply with sustainability standards (Macdonald, 2007). Some of the key challenges can be overcome by using blockchain-based institutional governance mechanisms. For instance, in Circulor's system, the details of the materials are entered by ASMs by registering on the system (Bennett, 2019). In Circulor's mine-to-manufacturer traceability solution for Tantalum, the enterprise application has two interfaces: (a) Mobile apps for checking IDs, scanning QR codes at checkpoints, and downloading documents; (b) Desktop versions for corporate offices to provide supply chain visibility and provide answers to queries. The system databases are hosted in Oracle Cloud and Amazon Web Services (Hyperledger, 2019). Specifically, ASMs use a mobile app, which is free for small companies and is easy to use (Bennett, 2019) whereas companies further up the supply chain need to pay and use more complicated interfaces.

The process begins with facial recognition. Once the miners open the app, there are three buttons on the front page. Using Circulor's system small mining companies do not see an increase in their workload. They may not know they are using blockchain. The final step of that registration is that the regulators approve it. Likewise, GemFair provides a tablet for a participating mine site to log in the GemFair app (Hill, 2018).

In this way, ASMs have access to relevant resources and competencies, which prior researchers have found to influence their likelihood to comply with sustainability standards. Moreover, weak and peripheral actors such as ASMs can influence information flows, which would make it difficult to conceal unsustainable and irresponsible practices. The most important of all is that the marginal groups such as miners are a part of the network. These groups thus get a voice and are represented in decision-making processes, which is a key step in the institutional transformation processes (Macdonald, 2007).

5.5   Chapter Summary and Conclusion

Non-blockchain supply chain systems suffer from a number of drawbacks that can be overcome by using blockchain systems, which can drastically improve the ability of firms to *identify, trace and track* information about people, organizations and materials. Blockchain-based systems provide authentic and reliable information to select exchange partners more effectively. Such information is also helpful to design incentive structures for supply chain partners that comply with sustainability-related expectations. *Compared to established traceability programs such as* ITSCI's: "bagging and tagging" system, blockchain solutions launched by some start-ups to trace minerals are more cost-effective.

Blockchain facilitates decentralized information flow which *reduces the prominence of powerful* actors in a supply chain. *Blockchain is thus characterized by a low degree of centrality. In blockchain-based supply chain networks,* actors along the value chain are more interconnected and there is an increased sharing of information, which increases monitoring of organizations. Whereas non-blockchain supply chain networks are standalone and discrete, outside actors such as regulatory agencies are *embedded in* blockchain-based supply chain *networks.* Blockchain-based supply chain networks can increase the transparency of information, which can address issues related to the disempowerment of marginalized communities by allowing their participation in the network.

References

Acharya, V. (2019). *Oracle blockchain platform (OBP)—A driver in proliferating blockchain adoption*. Government Blockchain Association. https://www.gba global.org/oracle-blockchain-platform-obp-a-driver-in-proliferating-blockc hain-adoption/

Airhart, E. (2018). *Alternatives to cobalt, the blood diamond of batteries*. Wired. https://www.wired.com/story/alternatives-to-cobalt-the-blood-diamond-of-batteries/

Allen, D. W. E., Berg, C., Davidson, S., Novak, M., & Potts, J. (2019). International policy coordination for blockchain supply chains. *Asia & the Pacific Policy Studies, 6*(3), 367–380.

Allison, I. (2020). *Everledger looks beyond blood diamonds with ESG supply chain collaboration*. Coindesk. https://www.coindesk.com/everledger-looks-beyond-blood-diamonds-with-esg-supply-chain-collaboration

Ayogu, M., & Lewis, Z. (2011). *Conflict minerals: An assessment of the Dodd-Frank Act*. Brookings. https://www.brookings.edu/opinions/conflict-minerals-an-assessment-of-the-dodd-frank-act/

Bates, R. (2018). *Inside Tracr, the De Beers–developed blockchain platform*. JCK. https://www.jckonline.com/editorial-article/tracr-de-beers-blockchain-platform/

Baydakova, A. (2019, January 16). *Ford, LG to Pilot IBM blockchain in fight against child labor*. https://www.coindesk.com/ford-lg-to-pilot-ibm-blockc hain-in-fight-against-child-labor

Bennett, M. (2019). *Blockchain app for miners in Rwanda ensures the minerals in your iPhone are conflict-free*. Diginomica. https://diginomica.com/blockc hain-app-for-miners-in-rwanda-ensures-the-minerals-in-your-iphone-are-con flict-free

Berg, C., Davidson, S., & Potts, J. (2018). *Ledgers*. https://papers.ssrn.com/sol3/papers.cfm?abstract_id=3157421

Blühdorn, I. (2007). Sustaining the unsustainable: Symbolic politics and the politics of simulation. *Environmental Politics, 16*, 251–275.

Bohlsen, M. (2019). *A look at the top 5 lithium-ion battery manufacturers in 2019*. Seeking Alpha. https://seekingalpha.com/article/4289626-look-top-5-lithium-ion-battery-manufacturers-in-2019

Boström, M., Jönsson, A. M., Lockie, S., Mol, A. P. J., & Oosterveer, P. (2015). Sustainable and responsible supply chain governance: Challenges and opportunities. *Journal of Cleaner Production, 107*, 1–7.

Boström, M. (2015). Between monitoring and trust: Commitment to extended upstream responsibility. *Journal of Business Ethics, 131*(1), 239–255.

Brugger, F. (2019). *Blockchain is great, but it can't solve everything. Take conflict minerals*. African Arguments. https://africanarguments.org/2019/04/23/blockchain-is-great-but-it-cant-solve-everything-take-conflict-minerals/

Burbidge, M. (2019). *Proving provenance*. Brainstorm Magazine. http://www.brainstormmag.co.za/business/14571-proving-provenance

Burns, E. (2016, September 29). *Private blockchains—Faster and more efficient*. Tech Monitor. https://techmonitor.ai/leadership/digital-transformation/private-blockchains-faster-efficient

Bush, S. R., Oosterveer, P., Bailey, M., & Mol, A. P. J. (2015). Sustainable chain governance: A review and future outlook. *Journal of Cleaner Production, 107*, 8–19.

Camerer, C., & Thaler, R. (1995). Ultimatum games. *Journal of Economic Perspectives, 9*, 209–220.

Cant, J. (2019). *Block firm helps Congo mine fight against blood diamonds*. Cointelegraph. https://cointelegraph.com/news/blockchain-firm-helps-congo-mine-to-fight-against-blood-diamonds

Cholteeva, Y. (2019). *Everledger launches blockchain platform to ensure transparency in diamond sourcing*. Mining Technology. https://www.mining-technology.com/news/everledger-launches-blockchain-platform-to-ensure-transparency-in-diamond-sourcing/

Cole, R., Stevenson, M., & Aitken, J. (2019). Blockchain technology: Implications for operations and supply chain management. *Supply Chain Management, 24*(4), 469–483.

Côme Mugisha, E. (2019). *Rwanda keen on accelerating development of blockchain*. The New Times. https://www.newtimes.co.rw/news/rwanda-keen-accelerating-development-blockchain

Debter, L., Dent, M., Del Castillo, M., Hansen, S., Kauflin, J., Sorvino, C., & Tucker, H. (2020). *Blockchain 50*. Forbes. https://www.forbes.com/sites/michaeldelcastillo/2020/02/19/blockchain-50/#770de3b67553

Devanesan, J. (2020). *Ford teams with IBM to trace cobalt on the blockchain*. TechHQ. https://techhq.com/2020/03/ford-teams-with-ibm-to-trace-cobalt-on-the-blockchain/

European Union. (2020). *Study on due diligence requirements through the supply chain*. Final Report. https://www.business-humanrights.org/sites/default/files/documents/DS0120017ENN.en_.pdf

Forbes Africa. (2018). *Blood diamonds to blockchain diamonds?* Forbes Africa. https://www.forbesafrica.com/technology/2018/09/03/blood-diamonds-to-blockchain-diamonds/

Forster, P. W., & Regan, A. C. (2001). Electronic integration in the air cargo industry: An information processing model of on-time performance. *Transportation Journal, 40*(4), 46–61.

Francisco, K., & Swanson, D. (2018). The supply chain has no clothes: Technology adoption of blockchain for supply chain transparency. *Logistics, 2*(1), 2.

Gemfair. (2019). *Artisanal and small-scale mining standard*. Gemfair. https://gemfair.com/static/files/GemFair_ASM_Requirements_2019_v2.pdf

Gleeson, D. (2019). *Volkswagen signs up blockchain specialist Minespider to track lead supply*. International Mining. https://im-mining.com/2019/04/25/volkswagen-signs-up-blockchain-specialist-minespider-to-track-lead-supply/

Graafland, J. J. (2002). Sourcing ethics in the textile sector: The case of C&A. *Business Ethics: A European Review, 11*(3), 282–294.

Hall, M. (2019, August 2). *The next generation of electric cars verified by blockchain*. https://blogs.oracle.com/blockchain/post/the-next-generation-of-electric-cars-verified-by-blockchain

Heide, J. B. (1994). Interorganizational governance in marketing channels. *Journal of Marketing, 58*, 71–85.

Hyperledger. (2019). *Circulor achieves first-ever mine-to-manufacturer traceability of a conflict mineral with Hyperledger Fabric*. Hyperledger. https://www.hyperledger.org/learn/publications/tantalum-case-study

IDEX. (2018). *Everledger announces the industry diamond time-lapse protocol*. IDEX. http://www.idexonline.com/FullArticle?Id=43757

Jamasmie, C. (2019). *De Beers expands pilot scheme in Sierra Leone to sell ethically sourced diamonds*. Mining.com. https://www.mining.com/de-beers-expands-pilot-scheme-sierra-leone-sell-ethically-sourced-diamonds/

Kapilkov, M. (2020a). *Volvo invests in blockchain startup to trace cobalt in its batteries*. Cointelegraph. https://cointelegraph.com/news/volvo-invests-in-blockchain-startup-to-trace-cobalt-in-its-batteries

Kapilkov, M. (2020b). *Startup helps reduce child labor in Africa & Aspires to work with Tesla*. Cointelegraph. https://cointelegraph.com/news/startup-helps-reduce-child-labor-in-africa-aspires-to-work-with-tesla

Khatri, Y. (2019). *Volkswagen to track minerals supply chains using IBM blockchain*. https://www.coindesk.com/volkswagen-to-track-minerals-supply-chains-using-ibm-blockchain

Kim, H. M., & Laskowski, M. (2018). Toward an ontology-driven blockchain design for supply-chain provenance. *Intelligent Systems in Accounting, Finance and Management, 25*(1), 8–27.

Kshetri, N. (2018). Blockchain's roles in meeting key supply chain management objectives. *International Journal of Information Management, 39*, 80–89.

Kshetri, N. (2020). *Blockchain's potential impacts on supply chain sustainability in developing countries*. Academy of Management Best Paper Proceedings, 7–11 August.

Lèbre, É., Stringer, M., Svobodova, K., Owen, J. R., Kemp, D. Côte, C., Arratia-Solar, A., & Valenta, R. K. (2020). The social and environmental complexities of extracting energy transition metals. *Nature Communications, 11*, Article number 4823. https://www.nature.com/articles/s41467-020-18661-9

LeVine, S. (2018). *A push for a battery leap to eliminate "blood cobalt".* Axios. https://www.axios.com/lithium-ion-batteries-blood-cobalt-drc-congo-49323694-40b9-4ff5-bbb2-fb0bb5d04bbe.html

Lewis, B. (2019). *Ford and IBM among quartet in Congo cobalt blockchain project.* Reuters. https://www.reuters.com/article/us-blockchain-congo-cobalt-electric/ford-and-ibm-among-quartet-in-congo-cobalt-blockchain-project-idUSKCN1PA0C8

London Mining Network. (2019, September 16). *New report exposes mining industry's greenwashing.* https://londonminingnetwork.org/2019/09/new-report-exposes-mining-industrys-greenwashing/

Luzzini, D., Caniato, F., & Spina, G. (2014). Designing vendor evaluation systems: An empirical analysis. *Journal of Purchasing and Supply Management, 20*(2), 113–129.

Macdonald, K. (2007). Globalising justice within coffee supply chains? Fair trade, Starbucks and the transformation of supply chain governance. *Third World Quarterly, 28*(4), 793–812.

MENA Report. (2019, November 11). *Germany: Volvo cars joins responsible sourcing blockchain network, Launched by IBM, Ford, and Volkswagen Group. Advancing ethical sourcing of minerals continues to scale with this network.* MENA Report.

Min, H. (2019). Blockchain technology for enhancing supply chain resilience. *Business Horizons, 62*(1), 35–45.

Mol, A. P. J. (2015). Transparency and value chain sustainability. *Journal of Cleaner Production, 107*, 154–161.

Narayanaswami, C., Nooyi, R., & Govindaswamy, S. (2019). Blockchain anchored supply chain automation. *IBM, 63*(2), 7:1–7:11.

Nelson, D. (2019, November 6). *Story from markets IBM ethical mineral sourcing blockchain to debut in spring.* https://www.coindesk.com/ibm-ethical-mineral-sourcing-blockchain-to-debut-in-spring

Neville, B. A., & Menguc, B. (2006). Stakeholder multiplicity: Toward an understanding of the interactions between stakeholders. *Journal of Business Ethics, 66*(4), 377–391.

Nkulu, B., Casas, L., Haufroid, V., De Putter, T., Saenen, N. D., Kayembe-Kitenge, T., Musa Obadia, P., Kyanika Wa Mukoma, D., Lunda Ilunga, J. M., Nawrot, T. S., & Luboya Numbi, O. (2018). Sustainability of artisanal mining of cobalt in DR Congo. *Nature Sustainability, 1*, 495–504. https://doi.org/10.1038/s41893-018-0139-4 https://www.nature.com/articles/s41893-018-0139-4?spm=smpc.content.content.1.1550534400023WRl5Apr

Norris, P. (2000). Global governance and cosmopolitan citizens. In J. S. Nye & J. D. Donahue (Eds.), *Governance in a globalizing world.* Brookings Institution Press.

rcsglobal.com. (2019). *Volvo cars joins responsible sourcing blockchain network launched by RCS Global, IBM, Ford, and Volkswagen Group*. RCS Global Group. https://www.rcsglobal.com/volvo-cars-joins-responsible-sourcing-blockchain-network-launched-rcs-global-ibm-ford-volkswagen-group/

Rolander, N. (2019). *Volvo cars goes for blockchain tech to avoid unethical cobalt*. Bloomberg. https://www.bloomberg.com/professional/blog/volvo-cars-goes-for-blockchain-tech-to-avoid-unethical-cobalt/

Ross, A., & Lewis, B. (2019). *Congo mine deploys digital weapons in fight against conflict minerals*. Reuters. https://www.reuters.com/article/us-congo-mining-insight/congo-mine-deploys-digital-weapons-in-fight-against-conflict-minerals-idUSKBN1WG2W1

Rowley, T. J. (1997). Moving beyond dyadic ties: A network theory of stakeholder influences. *Academy of Management Journal, 22*(4), 887–910.

Sabine, P. (2019). *How does blockchain root out blood diamonds from the world's supply market?* South China Morning Post. https://www.scmp.com/magazines/style/luxury/article/3031368/how-does-blockchain-root-out-blood-diamonds-worlds-supply

Sanderson, K. (2015). *Concerns raised over tantalum mining*. Nature. https://www.nature.com/news/concerns-raised-over-tantalum-mining-1.19023

Sedex. (2013). *Briefing transparency*. Sedex. https://www.sedex.com/briefing-transparency/

Taylor, P. (2020). *Volvo invests in cobalt traceability partner Circulor*. Securing Industry. https://www.securingindustry.com/electronics-and-industrial/volvo-invests-in-cobalt-traceability-partner-circulor/s105/a11996/#.XxzqrJ5KjIU

Thibodeaux, W. (2018). *A.I. is awesome, blockchain is a powerhouse. But here's what combining them could do*. Inc. https://www.inc.com/wanda-thibodeaux/ai-is-awesome-blockchain-is-a-powerhouse-but-heres-what-combining-them-could-do.html

Thompson, F. (2020). *Trade finance blockchain consortia: Where are we now?* Fintech. https://www.gtreview.com/magazine/volume-18-issue-2/trade-finance-blockchain-consortia-now/

Todd, F. (2019, October 1). *Human rights abuses and a Chinese monopoly: Can blockchain solve problems facing the DRC's mining sector?* https://www.nsenergybusiness.com/features/scanners-barcodes-ethical-mining-drc/

UNSDSN. (2016). *Mapping mining to the sustainable development goals: An Atlas (Columbia Center on Sustainable Investment)*. United Nations Development Program, UN Sustainable Development Solutions Network, World Economic Forum.

usgs.gov. (2020). *Cobalt statistics and information*. USGS. https://www.usgs.gov/centers/nmic/cobalt-statistics-and-information#:~:text=On%20a%20global%20basis%2C%20the,another%20major%20use%20for%20cobalt

Uwiringiyimana, C., & Lewis, B. (2018). *Wooing investors, Rwanda hosts first tantalum-tracking blockchain*. Reuters. https://www.reuters.com/article/us-rwanda-blockchain/wooing-investors-rwanda-hosts-first-tantalum-tracking-blockchain-idUSKCN1MR2ZQ

Volvo Cars. (2020). *Volvo Cars Tech Fund invests in blockchain technology firm Circulor*. Volvo Cars. https://www.media.volvocars.com/global/en-gb/media/pressreleases/269598/volvo-cars-tech-fund-invests-in-blockchain-technology-firm-circulor

War on Want and London Mining Network. (2019). *A just transition is a post-extractive transition*. https://londonminingnetwork.org/wp-content/uploads/2019/09/Post-Extractivist-Transition-report-2MB.pdf

Williamson, O. E. (1983). Credible commitments: Using hostages to support exchange. *American Economic Review, 73*(September), 519–540.

Wolfson, R. (2019). *Volvo Adopts Oracle's blockchain for its supply chain—Here's why*. CoinTelegraph. https://cointelegraph.com/news/volvo-adopts-oracles-blockchain-for-its-supply-chain-heres-why

Zucker, L. (1986). Production of trust: Institutional sources of economic structure 1840–1920. *Research in Organizational Behavior, 8*, 53–111.

CHAPTER 6

Discussion, Conclusion, and Recommendations

Abstract In this final chapter, we integrate the ideas discussed in earlier chapters regarding blockchain's potential to bring economic, political, and social transformations in the Global South. We consider the potentially transformative impacts of blockchain in diverse industries and markets in the Global South. The chapter also analyzes the barriers and challenges in implementing blockchain projects in the Global South. Finally, it looks into the future of blockchain in the Global South.

Keywords Degree of digitization · Economic transformation · Political transformation · Regulatory incompatibility · Social transformation · Standardization

6.1 INTRODUCTION

Blockchain offers a number of benefits to individuals and organizations in GS economies. For instance, blockchain can stimulate entrepreneurship in these economies. For small businesses, blockchain can help build trust which could be essential to get a loan from a bank. It can also increase the chances of finding a market for a service or product created in these economies.

N. Kshetri, *Blockchain in the Global South*,
https://doi.org/10.1007/978-3-031-33944-8_6

Fraud cases such as those in Qingdao are illustrative of illegal and unethical practices in GS economies. Technological intermediation by blockchain' (UNCTAD, 2020) provides a major opportunity for fighting corruption and fraud and reducing economic crimes such as embezzlement. Perpetrators find it difficult to engage in such unethical and illegal conducts because the transaction process is transparent.

An encouraging trend is that competition in the market has been increasing for blockchain solutions. Looking at the government administrative services, for instance, there are a number of initiatives that have attempted to develop blockchain solutions to drive efficiency in customs administration and facilitate trade. In November 2018, nine ocean carriers and terminal operators—COSCO Shipping Lines (China), Compagnie Maritime d'Affrètement and Compagnie Générale Maritime (CMA CGM), Evergreen Marine, Hong Kong-based Orient Overseas Container Line (OOCL), Yang Ming, DP World, Hutchison Ports, PSA International and Shanghai International Port, and CargoSmart—announced that they would form a consortium to develop a blockchain-based platform, Global Shipping Business Network (GSBN). The blockchain software will be created by CargoSmart, which is a software company funded by Hong Kong-based container shipping and logistics service company OOCL (Cosgrove, 2018). OOCL is a founding member of GSBN. Likewise, in early 2018, it was reported that AB InBev, Accenture, APL, Kuehne + Nagel, and a European customs organization tested a blockchain solution to exchange documents (Kapadia, 2018). While these companies are mainly based in developed countries, some of them have signficant operations in the developing world. In addition, we discussed above some blockchain-based solutions used in international trades in which most of the participants and beneficiaries are developing world-based.

The national and international spillover and externality effects arising from governments' adoption of blockchain also deserve mention. In Peru, news regarding Peru Compras' decision to use blockchain to fight corruption also generated interest in blockchain among local companies. Businesses can use blockchain to improve efficiencies in administrative processes. Stamping.io was also reported to be expanding into other Latin American countries Chile, Ecuador, and Colombia to help businesses in these countries improve business practices (Antonio Lanz, 2019). Allocation of resources to control corruption can also generate spillover effects in terms of reducing corruption in neighboring jurisdictions (Goel & Nelson, 2007).

6.2 POTENTIALLY TRANSFORMATIVE IMPACTS OF BLOCKCHAIN IN THE GS

There are a number of potential transformative effects of blockchain in the GS.

Reducing the Costs of Remittances

One estimate suggested that officially recorded remittance flows to low- and middle-income countries (LMICs) amounted $630 billion in 2022 (worldbank.org, 2022). The transaction costs of remittances, especially small ones, are very high. Immigrants use. In 2022Q2, the average cost of sending a $200 remittance to an LMIC was 6% using an officially approved route (Broom, 2023).

Transfer services such as Western Union cost as much as 7% of the transfer amount. To transfer 300 rands from South Africa to neighboring countries, transfer fees varied from 35 to 68.2 rand by bank draft, from 19.2 to 62.5 rands by electronic transfer, and from 25.3 rands by Money-Gram and 6.2 rands by iKobo's services. For small businesses involved in international trade, the costs of acquiring major international currencies such as dollars and euros are high (Kshetri, 2023).

Blockchain is fundamentally transforming the international remittance market as crypto remittances are being used to address challenges related to high costs and inefficiency. As an example, Ripple's On-Demand Liquidity leverages XRP (the payment network's cryptocurrency) to send money faster and at a lower fee. Some users of RippleNet in Global South economies include Vietnam's TPBank, Pakistan's Faysal Bank, the National Bank of Egypt, and Thailand's Siam Commercial Bank. The Siam Commercial Bank has teamed up with the digital money-transfer provider Azimo to use RippleNet. Using non-blockchain solutions, settling a remittance sent from Europe to Thailand takes more than one business day. With RippleNet, Siam Commercial Bank clears pounds and euros into Thai baht in less than a minute. Using on-demand liquidity, banks can avoid pre-funding, which allows them to settle remittances quickly. On-demand liquidity is especially attractive to payment companies and nonbanking institutions that are required to open overseas accounts to fund their transfers. Many such institutions face difficulties opening overseas bank accounts due to concerns related to money laundering (Kshetri, 2023).

In August 2021, Bitcoin exchange LocalBitcoins announced that there would be no deposit fees and transaction fees between wallets on the platform. The platform does not deal with fiat currency itself. Users can use the platform to transact with each other (Namcios, 2021). For instance, to send money from Venezuela to family members in Colombia, the sender buys bitcoin with Venezuelan bolivars using a bank transfer in LocalBitcoins. The sender then can search for sell offers of bitcoin in Colombia and choose the offer with the best exchange rate. After the seller of bitcoin in Colombia transfers Colombian peso to the family member's bank account, the sender transfers the bitcoin. A user reported that the whole process takes less than an hour. The platform charges a 1% fee to the user who offered to sell bitcoins (localbitcoins.com. n.d.). Other platforms such as Binance P2P and LocalCryptos offer similar services (CoinDesk, 2020).

In November 2019, Thailand's Siam Commercial Bank announced that it launched an AI-based robo-adviser to manage investment portfolios. The robo-adviser makes decisions for investors based on market conditions and their risk preferences. Consumers can start using the service with just US$100 in initial investment. By 2019, the bank had invested US$1.3 billion in AI, digital platforms, and other technology as part of its four-year capital spending program (Nguyen, 2019). Siam Commercial Bank expected that the new system would result in 100,000 new accounts by 2020 (Faridi, 2020a, 2020b).

As another example of a low-cost, fast remittance system based on blockchain, in 2018, Ant Financial introduced a blockchain-based cross-border remittance system. AlipayHK and the Philippines-based mobile money company GCash teamed up to offer real-time money transfer between Hong Kong and the Philippines, with significantly lower fees and higher speed and efficiency than traditional transfer services (Fast company n.d.). Standard Chartered Bank was part of the initiative. Customers make a few clicks with AlipayHK and the money reaches the GCash user's account in seconds. When a user submits a remittance application, all network participants, including AlipayHK, GCash, and Standard Chartered Bank, get a notification. The sender and receiver can track the money during the entire process (PYMNTS, 2018)[3].

Likewise, as of April 2021, BitPesa, which provides cryptocurrency-based remittances for five currencies across Africa, had transacted $235 million. By that time, it served more than 26,000 customers, compared to 6,000 in 2017. Western Union teamed up with Coins.ph in the Philippines to offer mobile wallets to consumers, which allow them to hold and spend with both local and cryptocurrencies. The partnership is designed to make it easier for consumers in the country to receive cash remittances (Webber, 2021).

Administrative Efficiency-Enhancing Mechanisms

Blockchain's use in administrative procedure has various efficiency-enhancing mechanisms. In many GS countries, international trade procedures are highly inefficient. Most administrative processes are handled manually. A main advantage of blockchain is to support the issuance of unfalsifiable electronic documents (UNCTAD, 2020). Blockchain-based solutions have already been implemented in many GS countries to facilitate this process.

Blockchain can also make it possible for national governments to delegate the implementation of specific policies and targets to sub-state/non-state actors. Administrative efficiency can be increased by allowing the private sector to perform some of the key functions. As noted above, a key area of application of blockchain in many GS countries is in the land registry. Prior researchers have found that the security of property rights is positively related to public expenditure efficiency (Afonso et al., 2009). Thus blockchain-based land registry not only enhances efficiency in activities such as buying and selling lands but also in public expenditure.

Circumventing Government Control and Censorship

Blockchain has the ability to circumvent government control and censorship. For instance, as discussed in Chapter 4, blockchain helped Chinese Internet users fight against the country's strict Internet censorship. Likewise, during the police brutality protests in Nigeria in October 2020, the government tried to block the protesters from using local payment platforms. The protesters thus faced barriers in collecting donations in the local currency. They switched to Bitcoin. In about a week, US$400,000 was raised by the protestors, of which Bitcoin accounted for about 40% (Ekpu, 2020). In the first forty days of 2021 the movement initiated by

Russian opposition leader and anti-corruption activist Alexei Navalny's received bitcoin donations worth $300,000, which was more than the total amount raised in the cryptocurrency in 2020 (Reuters, 2021). These factors may prompt some governments to enact regulations and carry out enforcement against cryptocurrencies.

Securing Data Stored in Government IT Systems

A key responsibility of governments in the digital era is to protect citizens' information from hackers and cybercriminals. Centralized databases are susceptible to hacking. In 2018, India's Bhoomi system experienced a security breach in which nefarious actors transferred 19 acres of government wasteland in Devanahalli taluk to a private individual. Some of the largest owners of the land, known as land sharks, were suspected to manipulate the records. the Bhoomi software had been breached twice before, in which hackers made attempts to transfer government properties to private persons (Akshatha, 2018).

It is not that blockchain-based systems do not possess any security risks. for instance, nefarious actors can also exploit loopholes in smart contract algorithms (Lemieux, 2017). Despite this, blockchain systems in general are considered to be more secure than centralized data management systems (Kshetri, 2021). For instance, even if a hacker is able to penetrate a network and change land records, multiple redundant and identical copies of the same records are stored in multiple computers. If one is breached, there are many others as backups. That is, data in the blockchain are distributed across many interlocked computers (Kshetri, 2021).

Access to Market for Marginalized Entrepreneurs

Entrepreneurs in the GS face challenges in accessing the market, finance, and other key factors of production. Prior researchers have noted the potential of digital technologies to democratize entrepreneurship. Among key mechanisms, they can provide access to international market knowledge and facilitate interactions with value chain partners (Pergelova et al., 2019). For instance, Indonesian farmers use mobile services to access market information related to product reviews and prices of crop (West, 2012). Likewise, in Trinidad and Tobago, workers in the fishing industry use cellphones to access market information such as prices for various fishes in various markets. Mobile apps also increase efficiency of their job (Mallalieu & Lessey, 2012). However, such interactions do not necessarily lead to economic transactions such as product sales. Blockchain makes it possible to perform proper evaluations of value chain partners and facilitate interactions that are trustworthy.

Blockchain can facilitate access to the market for disadvantaged entrepreneurs. For instance, in Africa, women are less likely to engage in entrepreneurship compared with their male counterparts (Bardasi et al., 2007). In the non-blockchain world, buyers who want to support local disadvantaged groups such as women farmers often are not in a position to do so due to the lack of authentic information. Such a challenge can be overcome by blockchain applications. For instance, Bext360's app determines the identity of the person selling the products.

In many blockchain solutions, vulnerable groups such as farmers and miners enter data related to their products in the ledger. This can facilitate their entrepreneurial activities in many ways. Prior research has noted that marginalized groups must be given a voice and represented in decision-making processes so that they can exercise some control over the institutional transformation processes (Macdonald, 2007). The Bext360 example illustrates that coffee farmers are able to take more control over supply chain processes thanks to blockchain-based solutions. Moreover, since what is measured is managed, the systems that allow small farmers to enter crop information are likely to create awareness about the quality of crops. In this way, blockchain-based solutions can raise farmers' consciousness regarding quality of their produce.

Increasing Transparency and Accountability Among Government Agencies and Fighting Corrupt Practices

Blockchain can help governments to undertake their responsibilities more effectively. As noted above, Latin America's governments have been accused of abandoning their responsibility to fight poverty (Fox, 1994). In light of the accusations, it is encouraging that some governments in the region are using blockchain in order to discharge their responsibility to fight poverty. The government of Argentina has teamed up with NEC, the Innovation Laboratory of the Inter-American Development Bank

Group IDB Lab, and NGO Bitcoin Argentina to launch a blockchain-based digital identity for inclusion project in order to improve citizens' access to governmental services. A main component of the project is to deploy a blockchain-based ID system in Buenos Aires to improve access to quality goods and services. According to the Buenos Aires City Government (GCBA), 16.2% of the city's population lives below the poverty line. The concept of "poverty penalty is used to describe the levels of economic challenges faced by this population. The idea is that the poor pay relatively higher costs to access certain goods and services. Among the major causes of the poverty penalty is the existence of imperfect information. Due to the lack of information regarding the identity and behavior of the economically vulnerable group, the market tends to exclude them or include them at higher costs than those paid by the average population (NEC Corporation, 2019).

The absence or lack of morality among government officials has been a concern in many developing countries. Blockchain-based solutions can decrease the tendency to engage in immoral and corrupt practices since such actions are noticed by other participants in the networks. Several developing countries are exploring blockchain for corruption-prone activities such as public procurement.

In public procurements, for instance, blockchain-based solutions ensure that selection criteria are not tailored to favor specific contractors after the RFP is published and provide permanent and tamper-proof bid records. They can also allow the public to monitor actions and decisions related to a procurement process. All vendors can participate in the procurement process. Thus, restrictive corruption practices are controlled. As noted above, the Bhoomi system in India's Karnataka state increased corruption by centralizing land records and moving the management moved away from the village. This is a form of expansive corruption (Osterfeld, 1992). Blockchain-based land registries make it possible to complete land-related transactions efficiently without paying bribes, which reduces expansive corruption.

The World Bank has provided several recommendations aimed at addressing corruption. First, it is important to create a political will and constructive pressure. International organizations and donors should provide assistance in fighting corruption. They include supporting government reforms in economic policies and institutions. They can also play a valuable role in developing a healthy civil society (Gray & Kaufman, 1998). Prior researchers have emphasized the importance of addressing accountability and transparency in government procurement systems in developing countries (Raymond, 2008).

The above discussion has shown that many of these goals can be accomplished by using blockchain. Prior researchers have emphasized the importance of combining formal and informal accountability to enhance network performance (Romzek et al., 2012). Blockchain systems can be designed in a way that allows the sharing of data and information with many participants so that responsiveness and accountability can be ensured. For instance, blockchain provides additional channels of informal accountability, such as the participation of teachers and parents in school meal programs.

6.3   Barriers and Challenges

Despite blockchain's potential noted in the previous section, its adoption has a number of major challenges to overcome. In this section, we explore some of the key barriers and challenges that deter organizations from adopting blockchain in the GS economies.

The Lack of Institutional Capacities

Blockchains can be viewed as a technology that governs information and exchange (Davidson et al., 2018). Note that a governance system is the "totality of institutional arrangements—including rules and rule-making agents—that regulate transactions inside and across the boundaries of an economic system" (Hollingsworth et al., 1994, p. 5). A key point that needs to be emphasized is that blockchain systems are embedded in the broader political and economic systems and they should not be viewed as a self-contained phenomenon with self-contained solutions. Blockchain's effectiveness in achieving effective SC governance thus depends on the nature of the power struggle between actors in the broader context of the political economy.

Global supply chains operate in a complex environment that requires various parties to comply with diverse laws, regulations, and institutions. They include maritime laws and regulations, commercial codes, and laws pertaining to ownership and possession of multiple jurisdictions in the shipping routes. Since international businesses operate against the backdrop of these established old laws, customs, and institutions that are managed by human beings, implementing blockchain-based solutions can be an extremely complex task (Casey & Wong, 2017). Addressing this challenge may be no small feat.

Especially political and institutional arrangements in many GS economies are among the most salient barriers that prevent the deployment of blockchain. Countries associated with conflict minerals are experiencing a high degree of corruption. In most parts of the DRC, for instance, the central government authority is virtually non-existent (McFerson, 2009). The DRC's social hierarchy system is described as neo-patrimonialism in which state resources are used by patrons to ensure the loyalty of clients in the general population (Emizet, 2000). The patron–client relationship, which is often informal, reaches from the state structures to small villages. The DRC is one of the most corrupt countries in Africa due to its economic resources, the lack of good governance, and its history of repression and corruption during as well as after colonialism (McFerson, 2009). In 2019, the DRC ranked 168 out of 180 countries in terms of the Corruption Perceptions Index. In the previous 12 months of the survey period, 85% of the country's people thought corruption increased and 80% of public service users needed to pay a bribe.

Since no one party has all data related to the ownership and flow of products in an SC, blockchain systems provide incentives for supply chain participants to provide information so that the provenance and state of products can be evaluated (Kim & Laskowski, 2018). However, some parties may have incentives to provide false information related to product provenance. In some institutional settings, there is no or insufficient penalty for providing false and incorrect information.

Low Degree of Digitization

Most serious environmental, social, and governance (ESG) risks reside deeper down in the supply chain (Sedex briefing, 2013). These include vulnerable smallholder farmers in developing economies that grow subsistence and cash crops and ASMs in Africa.

Due to the requirement of a high degree of computerization, not all countries are ready to participate in blockchain-based solutions. Many supply chain partners located in developing and least-developed countries often are far from ready to adopt blockchain because they face a significant challenge to digitize their supply chains (Fig. 6.1). Without their participation, it is difficult to realize the full potential of blockchain in the supply chain. For instance, as of 2015, mobile network coverage was estimated to

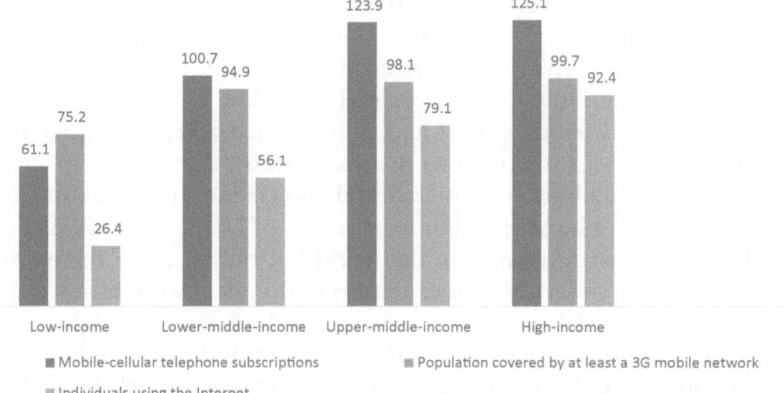

Fig. 6.1 Key indicators related to ICT access in economies with various levels of economic development (2022, percentage of the population). *Data source* International Telecommunications Union (https://www.itu.int/en/ITU-D/Sta tistics/Pages/stat/default.aspx)

reach 50 to 60% of the DRC population (Siedek, 2015). Likewise, significant proportions of the population in other mineral-rich African countries such as Burundi and Zambia lack mobile network coverage. In general, as of 2019, more than 21% of the population in the least developed countries (LDCs), which are low-income countries that perform poorly in human assets and face high economic vulnerability (United Nations, 2022) lacked mobile network coverage. Likewise, as of 2022, more than 38% of the population in low-income countries did not have a cellphone and more than 73% lacked Internet access (Fig. 6.1).

Lack of Technological Expertise and Absorptive Capacity

Although many companies describe themselves as "blockchain companies", few real use cases have emerged. We described numerous innovation-related activities in the blockchain arena that are taking place in China (Chapter 4). A caveat however is that patent count as a way to measure innovation has drawbacks. For instance, among China's 262 public companies that self-categorized themselves as "blockchain companies," as of September 2020, only 23 had mentioned blockchain use cases that had gone live (Kong, 2020). Despite the mushrooming of blockchain companies in China, most of them have started companies with extremely low investments. According to LongHash, 46% of the companies were registered with a capital of as low as 5,000 (US$717) yuan or less (Daily Hodl, 2020) and only 9% had a capital higher than 50,000 yuan (US$7,170) (Crypto Browser, 2020).

The lack of capabilities such as technical knowledge and skills also affect the deployment of blockchain. For instance, as of 2018, there were about 20 million software developers in the world but only 0.1% of them knew about blockchain codes. No more than 6,000 coders were estimated to have the levels of skill and experience needed to develop high-quality blockchain solutions (Suprunov, 2018). Out of India's 2 million software developers, only 5,000 were estimated to have blockchain skills. Some speculate that about 80% of these developers may pursue job opportunities outside the country (Agarwal, 2018).

It is unreasonable to expect that blockchain solutions can be sent to rural Africa for artisanal miners to use them (Early, 2019). Even if such systems are set up with outside help, small farmers, and miners cannot.

Cost-Related Barriers

Like other technologies, blockchain deployment tends to diffuse from larger to smaller organizations. This is also known as the rank effect (Gotz, 1999). Due to cost and complexity, blockchain systems are expensive to implement and manage. For this reason, blockchain is out of reach for many organizations. For instance, JD's supply chain partner Kerchin which has adopted blockchain had US$ 300 million in revenue in 2017 (Chapter 4). Chinese firms also face human resources challenges. As noted above, Walmart needed to train employees and suppliers to use its blockchain platform in China. Most of the food products in China are produced on very small farms that lack access to technology or Internet connectivity. This is the main reason why food safety has been difficult to achieve in China (McMillan, 2018). Most of these small farms do not possess the capability to adopt a blockchain-based system and provide relevant information.

Blockchain systems are thus expensive to implement and manage. Despite some smaller companies' access to blockchain-based solutions as noted above, due to cost and complexity, this technology is out of reach for many organizations. For instance, as of August 2020, the costs per month of using IBM enterprise blockchain platform, which is based on Hyperledger Fabric (devteam.space, 2020), included a membership fee of US$1000 and per-peer fee of US$1,000 (IBM, n.d.). This translates to an annual peer fee of US$12,000 for each additional member added to the IBM enterprise blockchain platform.

Regulatory Incompatibility and the Lack of Standardization

Past paperless trade processes mostly entailed digitizing existing forms and processes. Blockchain-led digitization is different in the sense that it enables new data structures. Blockchain data structures may also lead to regulatory incompatibility. That is, new forms of data in blockchains and distributed ledger technologies (DLTs) may not satisfy the regulatory requirements of some jurisdictions. The creation of standards for data structures is necessary for at least two reasons. First, from a technical perspective, different blockchain platforms may carry SC information and thus the interaction of data between different blockchains is critical. Second, looking from an information perspective, standards of data structure are needed to comply with different regulatory requirements (Allen et al., 2019).

Standardization and interoperability have been a concern for the use of enterprise permissioned blockchains (Faridi, 2020a, 2020b). For instance, in supply chains, different suppliers use different blockchains that lack interoperability (Buntinx, 2017). Likewise, in 2019, Euromoney reported that there were about 30 consortia that focused on using DLT in trade finance (euromoney.com, 2019). Each consortium consists of many different banks that use different platforms with slightly different offerings. After an organization joins a consortium, it is difficult to break the self-reinforcing mechanism associated with the consortium (Arthur, 1988).

Achieving Deep Collaboration and Partnership

Another key challenge that arises in the implementation of blockchain consists of bringing all the relevant parties together, which can be a difficult undertaking in many cases. Everledger Founder and CEO Leanne Kemp noted that it took about 18 months to negotiate the relationships needed to make the Everledger service possible (Clancy, 2017). Trade-Lens represents one of the best examples to illustrate this situation (In Focus 6.1).

Especially downstream firms' lack of willingness-to-pay for SC visibility has been a concern in the mineral sector. For instance, most ASMs in the DRC sell to the Société Minière de Bisunzu (SMB), which is authorized to buy minerals found in Rubaya, which is the capital of the North Kivu province and has the largest coltan mining site in DRC. SMB then sells the minerals to the international market. In an April 2019 report, the consulting group focusing on African markets Sofala Partners estimated that SMB buys between $3 million and US$5 million worth of coltan each month from about 3,500 artisanal miners in Rubaya (Desai et al., 2019). Citing data from the DRC's Ministry of Mines, news website Reuters reported that in 2018, SMB had supplied ore to two smelting Asian companies, which "are, or may be" in the supply chains of Tesla, GM, Ford and Apple according to their public filings(Desai et al., 2019). U.S. carmakers and technology companies have mostly exhibited indifferent attitude toward conflict minerals. Reuters reported that Tesla did not respond to requests for comment and Apple declined to comment. GM and Ford referred to their filings with the U.S. SEC (Desai et al., 2019).

In Focus 6.1: TradeLens shuts down its operation

In August 2018, Maersk and IBM announced that the two companies jointly developed a blockchain-powered shipping solution TradeLens (https://www.tradelens.com/). The goals of TradeLens are to bring various parties involved in international trade together, support information sharing among them and enhance transparency.

As soon as containers leave, customs authorities at the destination country would begin to receive shipping data. TradeLens also uses IoT devices to record physical aspects of shipping such as temperature, and container weight. Officials will have time to prepare documentation. There is also an immutable log of data such as where the ship has been. It facilitates auditing and fraud detection (Wood, 2020). TradeLens was reported to reduce waiting times for shipping documentation from days to minutes.

As of March 2020, TradeLens network consisted of 150 members. That included five of the world's top six ocean carriers– APM-Maersk, Mediterranean Shipping Company (MSC), China Ocean Shipping Company (COSCO), Hapag-Lloyd and Ocean Network Express (ONE) (Morris, 2019). Together they represent over half the world's container cargo capacity. By March 2020, the platform had processed 15 million containers (Ledger Insights, 2020). By November 2020, more than 175 organizations that covered more than 600 ports and terminals had joined. They posted more than 1.6 billion transport events involving 34 million containers to the blockchain and processed 14 million documents (Lacity & Van Hoek, 2021).

Government agencies from the Global South to join TradeLens include customs in Malaysia, Thailand, Saudi Arabia, and Jordan (Ledger Insights, 2020). In October 2020, Royal Malaysian Customs Department (RMCD) announced a plan to use TradeLens to modernize the shipping processes. The adoption of TradeLens is expected to increase security and transparency, enhance customer satisfaction and facilitate the ease of doing business. TradeLens will provide RMCD with an automatic and immutable tracking tool, a simpler workflow and near real-time information sharing from other TradeLens members (TradeLens, 2020).

Sri Lanka's South Asia Gateway Terminals (SAGT), which is the first Public Private Partnership (PPP) container terminal has collaborated with TradeLens. This is expected to digitize manual, paper-based and time-consuming administrative processes. As of July 2020, more than 110 ports and terminals, more than 15 customs authorities and many intermodal providers were using TradeLens (Loadstar, 2020).

In late 2022, A.P. Moller - Maersk and IBM announced that they would end TradeLens operations in the first quarter of 2023 (Cecere, 2022). A.P. Moller—Maersk noted that a "full global industry collaboration has not been achieved" (Maersk, 2022).

Solutions Often Favoring the Interests of Big Companies Rather Than Low-Income People

Many blockchain-based systems are designed to benefit big companies rather than disadvantaged groups. For instance, in the BanQu systems in India, Uganda, and Zambia to track cassava and barley, Anheuser-Busch can benefit tremendously from blockchain's use to promote SC transparency and traceability. Blockchain can help to guarantee the quality of products with relevant data. Making digital payments to farmers may lower the costs associated with payments. However, these big firms have done very little to ensure that farmers can genuinely benefit from the integration of blockchain in their supply chains.

For instance, smallholder farmers that supply crops to Anheuser-Busch may theoretically enjoy additional benefits (e.g., getting low-cost loans from financial institutions) using their identity and transaction information put on BanQu's blockchain. However, constraints related to information flows, transaction costs, and market access would prevent them from realizing such benefits. For instance, the farmers may not be able to present the information in a way that meets the requirement of banks. They may also lack persons in their social network who possess the capability to understand the various available loan services. Due to the lack of education, many potential borrowers cannot fill out loan applications (Kshetri, 2020). Poor people often need loans in small amounts. It is costly for financial institutions to deal with small transactions. In some cases, poor people may face prejudice and stereotypes. Some banks refuse their admission to bank branch offices (Thorsten et al., 2009).

In many blockchain-based solutions, one of the stated goals has been to improve the access to and affordability of financial services for low-income people. Instead, the startups are keeping big companies' and investors' needs and preferences as a higher priority than those in need. New users of such technologies often lack the skills and access to new opportunities that can be derived from such new technologies (Kshetri, 2020). For instance, big companies such as Anheuser-Busch can benefit tremendously from blockchain's use to promote supply chain transparency and traceability. Blockchain can help to guarantee the quality of products with relevant data. Making digital payments to farmers may lower the costs associated with payments. Blockchain is being used by some firms to enhance reputational value by demonstrating their ability to innovate and increasing consumers' perception of food safety (Higginson et al., 2019).

Due to increased pressures to meet stockholder expectations, large firms' philanthropy activities have been declining (Porter & Kramer 2002). This is the main reason why relatively few genuine efforts have been made to help disadvantaged groups engage in entrepreneurial activities. In many cases, the stated benefits of disadvantaged groups' participation in blockchain networks are conditional rather than guaranteed. For instance, smallholder farmers that supply crops to Anheuser-Busch may theoretically enjoy additional benefits (e.g., getting low-cost loans from financial institutions) using their identity and transaction information put on BanQu's blockchain. However, constraints related to information flows, transaction costs, and market access would prevent them from realizing such benefits. For instance, the farmers may not be able to present the information in a way that meets the requirement of banks. They may also lack persons in their social network who possess the capability to understand the various available loan services. Due to the lack of education, many potential borrowers cannot fill out loan applications. Poor people often need loans in small amounts. It is costly for financial institutions to deal with small transactions. In some cases, poor people may face prejudice and stereotypes. Some banks refuse their admission to bank branch offices.

Less Beneficial and Less Impactful Uses

Some high-profile projects have also been launched that have focused on less beneficial and less impactful uses of blockchain. One example is Akon's ambitious crypto projects launched in Senegal (In Focus 6.2).

Likewise, blockchain-based games are often viewed as economically unproductive or low-return activities. Some such games do not require access to advanced technological resources. In February 2022, MetaCelo was launched as the first P2E metaverse NFT Game on Celo Network (accesswire.com, 2022). The game is designed for low-end smartphones with slow connections to deal with the problems of high latency and low bandwidth (Insider Studios, 2022). However, observers argue that, in general, play to earn (P2E) games lack sound economic fundamentals. P2E models require constant user growth and are zero sum. No money is produced inside the game.

In Focus 6.2: Akon's Ambitious Crypto Projects in Senegal Show Poor Performance

In 2018, Senegalese-American singer Aliaune Damala Badara Akon Thiam (known as Akon) announced two ambitious projects to be launched in Senegal that were supposed to transform the country. The first project was a $6 billion city. The second was a cryptocurrency called Akoin.

As of 2023, both were facing difficulties and delays. The site where the city was proposed to be built was a waste ground. The Akoin cryptocurrency was launched on Bitmart in September 2021. Its value at that time was $0.28, which reduced to $0.01 in January 2023. The legal tender in Senegal is the CFA franc, which is regulated and issued by Central Bank of West African States (BCEAO). The institution warned that there were risks of adopting cryptocurrencies such as Akoin and declared that it was illegal (Griffin, 2022).

6.4   FUTURE PROSPECTS

Despite the above barriers, blockchain-based solutions' future potential to address economic and social challenges facing the GS is even greater. Solutions relying on blockchain and cryptocurrency are developing at a rapid pace. For instance, solutions are being developed in which if a mining company claims that living wages are being paid to its miners, the accuracy and truthfulness of such claims can be assessed by checking the payments to digital wallets that are assigned to the miner (Early, 2019). One company working on such a solution is the U.S.-based blockchain company BanQu utilizes blockchain to establish economic identities and proof of record (which it calls 'economic passports') for unbanked persons (Stanley, 2017). BanQu has developed such solutions that can be used by farmers in India, Uganda, and Zambia to track cassava and barley supplied to the subsidiaries of the multinational drink and brewing company Anheuser-Busch (Kshetri, 2021). As of 2019, BanQu was working with telecommunications companies, battery and smartphone manufacturers, and jewelers to develop similar solutions for the mineral and mining industry. The company's plan is to launch the solutions for cobalt mines in the DRC, Zambia or Madagascar, and precious metal or gemstone mines in Botswana, Peru, and Colombia. For instance, when firms in the cobalt supply chain add information to BanQu's blockchain, the miner will receive an SMS message, which confirms key data such as the quantity sold and the price. The SMS sent to the miner is also stored on the blockchain. This means that if a cobalt buyer has not paid the correct amount to the miner, the data would not match and the end-user of the mineral would know it. For the miner the SMS record serves many purposes including proof that they are a part of a legitimate global supply chain (Early, 2019).

Some technological solutions are available but not being utilized up to their potential to address sustainability issues in mineral and metal supply chains. One such example is the Analytical Fingerprint (AFP), which can be employed to check the documented origin of tin, tungsten, and tantalum (3 T) ore minerals and to ensure that smuggled minerals do not enter a supply chain. This technique involves comparing a sample from a shipment to reference samples stored in a database to test the claim regarding the documented origin of the mineral. AFP relies on the identification of geochemical features or the distribution of chemical elements in mineral deposits from a given geographic location. In this way AFP can evaluate the plausibility of a claim regarding the origin stated in the documents of a mineral's shipment (BGR, n.d.). Germany's central geoscientific authority that provides advice to the Federal Government in geo-relevant issues Federal Institute for Geosciences and Natural Resources [Bundesanstalt für Geowissenschaften und Rohstoffe] (BGR) started developing an analytical fingerprinting (AFP) method since 2006 (BGR, n.d.). The initiative was launched in response to calls by the UN for a scheme to verify the origin of conflict minerals mined in the DRC and neighboring countries. The BGR's recommendation is to apply AFP as an optional forensic tool to perform audits or risk assessments in the uppermost section of a mineral supply chain. Specifically, the BGR has suggested performing AFP after extracting minerals from the mine sites and before homogenizing, that is, mixing the minerals in order to reduce the variance of the product supplied, for loading in a container for export. AFP can serve as a way to verify the integrity, and credibility of other traceability schemes (BGR, n.d.).

Tantalum and Niobium (Ta-Nb), which belong to so-called transition metals due to their positions in the periodic table of elements, are almost always found paired together. These two metals are difficult to separate due to their shared physical and chemical properties. Specifically, columbite ore which is rich in niobium and tantalite ore, which is rich in tantalum form a solid solution series known as the Columbite-Tantalite series (Minerals.net, n.d. a). When intermediary minerals exist between two end-member minerals[1] that are isomorphous (Minerals.net, n.d. b) (that is, the same crystal form because of identical molecular arrangement in spite of their different specific elements) a solid solution series is formed (Cengage, 2020).

Ta-Nb ores are extremely complex in terms of mineralogical and chemical composition. This is because columbite-tantalite solid solution series found in different parts of the world vary widely (Melcher et al., 2008). Moreover, the columbite-tantalite solid solution series incorporates many additional elements. The wide variations in Ta-Nb minerals and ores also provide opportunities to develop mineralogical-geochemical-geochronological-based fingerprinting schemes in order to determine their origins. It was reported that as of 2008, over 350 samples of individual crystals and ore concentrates had been analyzed by BGR. More than 60% of them were from central and southern Africa (Melcher et al., 2008).

Melcher et al.'s (2008) used an electron microprobe (EMP), which is an analytical tool used to determine chemical compositions of small volumes of solid materials in a non-destructive way, to study minerals. The team found different mineralogical and geochemical fingerprints of minerals depending on the origin of a concentrate. For instance, bismutotantalite was found only in samples from Mozambique. Wodginite was frequently found in samples from Rwanda. Tapiolite was detected in concentrates from the DRC and Rwanda (Melcher et al., 2008). By utilizing AFP for minerals in the first mile of the supply chain, the reliability, and authenticity of information provided by the sellers of cobalt regarding the mineral's origin can be assessed.

---

[1] In an isomorphous (solid-solution) series, an endmember is one of the simple compounds (Definition of end member. *Mindat.org*; https://www.mindat.org/glossary/end_member

6.5 FINAL THOUGHT

Although it can be argued that it is in the interest of providers of blockchain-related services to exaggerate the potential benefits of this technology, the analysis of this book suggests that blockchain, in combination with other technologies, such as the IoT and cloud computing, can drive economic, social, and political transformations in developing economies.

Economies in the GS need to deal with various challenges and bottlenecks in successful deployment of blockchain. Powerful actors that are against transparency and openness may oppose blockchain. In the land ownership example, blockchain can increase the transparency of land ownership and records and make it difficult or impossible for corrupt officials to alter land registries after the records are on the blockchain. Nonetheless, blockchain cannot address corruption in decisions about how land is registered in the ledger.

While many bold claims have been made about the effectiveness of blockchain in addressing many socioeconomic challenges and supporting evidence so far has not been convincing and conclusive, this technology could be a game-changer if it is appropriately deployed to address various challenges facing the GS economies. Thus, much more needs to be done to stimulate the adoption of blockchain to bring positive economic and social change to the GS.

The low degree of digitization and the lack of institutional capacities hinder developing countries' efforts to benefit from blockchain. Achieving deep collaboration and partnership among various actors, which is required for the deployment of blockchain solutions, is not easy.

Just like mobile telephony or mobile banking, blockchain in these countries may present an opportunity to leapfrog legacy IT systems. Due to a low level of trust and the high price of intermediation in many developing countries, blockchain demonstrates a high-value proposition. Blockchain is, therefore, an important tool that can be used in replacing the need for institutional and personal intermediation. Widespread adoption of blockchain may also enhance a country's image. For instance, the Republic of Georgia has been promoting itself as a corruption free-country with modern and transparent governance.

References

accesswire.com 2022. MetaCelo—The New Star in Celo Ecosystem, , February 16, 2022, https://www.accesswire.com/689027/MetaCelo--The-New-Star-in-Celo-Ecosystem

Agarwal, M. (2018). Blockchain: India likely to see brain drain as 80% developers may move abroad. *Inc42*, https://inc42.com/buzz/blockchain-india-likely-to-suffer-brain-drain-as-80-developers-prepare-to-move-abroad/

Akshatha, M. (2018). Karnataka's famed land record database Bhoomi faces another security breach. *The Economic Times*, https://tech.economictimes.indiatimes.com/news/corporate/karnatakas-famed-land-record-database-bhoomi-faces-another-security-breach/65748534

Allen, D. W. E., Berg, C., Davidson, S., Novak, M., Potts, J. (2019). International policy coordination for blockchain supply chains. *Asia & the Pacific Policy Studies, 6*(3), 367–380.

Antonio Lanz, J. (2019). *Peru sets its eyes on blockchain to fight government corruption*. Decrypt, https://decrypt.co/6893/peru-blockchain-government-corruption

Arthur, B. W. (1988) Self-reinforcing mechanisms in economics. In P. W. Anderson, K. J. Arrow, & D. Pines (Eds.), *The economy as an evolving complex system*. Perseus Press.

Bardasi, E., Blackden, M. C., & Guzman, J. C. (2007). *Gender, entrepreneurship, and competitiveness in Africa*. Africa Competitiveness Report, Chapter 1.4. The World Bank.

BGR (n.d.). Introduction to the Analytical Fingerprint. BGR, https://www.bgr.bund.de/EN/Themen/Min_rohstoffe/CTC/Analytical-Fingerprint/analytical_fingerprint_node_en.html

Broom, D. (2023). Migrant workers sent home almost $800 billion in 2022. which countries are the biggest recipients? *World Economic Forum*, https://www.weforum.org/agenda/2023/02/remittances-money-world-bank/

Buntinx, J. P. (2017). Top 6 mistakes enterprises make when exploring blockchain technology. *The Merkle*, https://themerkle.com/top-6-mistakes-enterprises-make-when-exploring-blockchain-technology/

calbitcoins.com. (n.d.). *LocalBitcoins.com: Fastest and easiest way to buy and sell bitcoins—LocalBitcoins*. [online] https://localbitcoins.com/fee

Casey, M. J., & Wong, P. (2017) Global supply chains are about to get better, thanks to blockchain. *Harvard Business Review*, https://hbr.org/2017/03/global-supply-chains-are-about-to-get-better-thanks-to-blockchain

Cecere, L. (2022). TradeLens discontinues operations. Why you should care., Forbes. *Forbes Magazine*. https://www.forbes.com/sites/loracecere/2022/12/05/tradelens-discontinues-operations-why-you-should-care/?sh=4d40f41d4cec

Cengage. (2020). Solid Solution Series. *Encyclopedia.com*, https://www.enc yclopedia.com/science/encyclopedias-almanacs-transcripts-and-maps/solid-solution-series#:~:text=A%20solid%20solution%20series%20is,one%20or%20m ore%20atomic%20sites

Clancy, H. (2017). Unilever teams with big banks on blockchain for supply chain. *Greenbiz*, https://www.greenbiz.com/article/unilever-teams-big-banks-blockchain-supply-chain

CoinDesk. (2020). *Crypto remittances prove their worth in Latin America.* [online] https://www.coindesk.com/crypto-remittances-latin-america-geopol itical-tension

Cosgrove, E. (2018). 9 ocean carriers, terminal operators join new blockchain initiative to rival TradeLens. *SupplyChainDive*, https://www.supplychaindive. com/news/ocean-carriers-new-blockchain-cosco-cma-cgm/541630/

Crypto Browser. (2020). Over 10,000 new Chinese blockchain companies in 2020 despite Covid-19. *The Capital*, https://medium.com/the-capital/over-10-000-new-chinese-blockchain-companies-in-2020-despite-covid-19-e6a38f 5323ad

Daily Hodl. (2020). China's blockchain boom moves forward, with 10,000 new companies formed in 2020: Report. *Daily Hodl*, https://dailyhodl.com/ 2020/08/17/chinas-blockchain-boom-moves-forward-with-10000-new-com panies-formed-in-2020-report/

Davidson, S., De Filippi, P., & Potts, J. (2018). Blockchains and the economic institutions of capitalism. *Journal of Institutional Economics, 14*(4), 639–658.

Desai, P., Shabalala, Z., & Daly, T. (2019). Exclusive: London metal exchange to delay ban on tainted metal until 2025—sources. *Reuters*, https://www.reu ters.com/article/us-lme-metals-sourcing-exclusive/exclusive-london-metal-exchange-to-delay-ban-on-tainted-metal-until-2025-sources-idUSKBN1W 31V2

devteam.space. (2020). How much does it cost to build a blockchain project? *DevTeam.Space*, https://www.devteam.space/blog/how-much-does-it-cost-to-build-a-blockchain-project/

Early, C. (2019). Can high-tech solutions take the risk out of artisanal mining? *Reuters*, https://www.ethicalcorp.com/can-high-tech-solutions-take-risk-out-artisanal-mining

Ekpu, E. U. (2020, December 17). Nigeria is now the No.2 bitcoin market on this fast-growing global marketplace. *Quartz*, https://qz.com/africa/194 7769/nigeria-is-the-second-largest-bitcoin-market-after-the-us/

Emizet, K. N. F. (2000). Congo (Zaire): corruption, disintegration, and state failure. In E. W. Nafziger, F. Stewart, & R. Vayrynen (Eds.), *Weak states and vulnerable economies: Humanitarian emergencies in the third world*. Oxford University Press.

euromoney.com. (2019) Trade finance and blockchain: Now is the time for a network of networks. *Euro Money*, https://www.euromoney.com/article/b1h041crxm5dks/trade-finance-and-blockchain-now-is-the-time-for-a-net work-of-networks

Faridi, O. (2020a). Blockchain interoperability: World economic forum releases paper explaining importance of effective communication between DLT networks. *Crowdfund Insider*, https://www.crowdfundinsider.com/2020a/04/160011-blockchain-interoperability-world-economic-forum-releases-paper-explaining-importance-of-effective-communication-between-dlt-net works/

Faridi, O. (2020b). People's Bank of China acquires $4.7 million in funding to further develop blockchain-based trade finance platform. *Crowdfund Insider*, https://www.crowdfundinsider.com/2020b/03/158535-peoples-bank-of-china-acquires-4-7-million-in-funding-to-further-develop-blockchain-based-trade-finance-platform/

Fast company. (n. d.). Ant financial. *Fast Company*, https://www.fastcompany.com/company/ant-financial

Fox, J. (1994). Latin America's emerging local politics. *Journal of Democracy*, 5(2), 105–116.

Goel, R., & Nelson, M. (2007). Are corrupt acts contagious?: Evidence from the United States. *Journal of Policy Modeling, 29*(6), 839–850. https://www.sciencedirect.com/science/article/pii/S0161893807001056

Gray, C. W., & Kaufman, D. (1998). Corruption and development. PREM Notes; No. 4. World Bank©. https://openknowledge.worldbank.org/han dle/10986/11545 License: CC BY 3.0 IGO.

Griffin, J. (2022). Akon's Wakanda, grazing goats and a crumbling crypto dream, *BBC News*. BBC. https://www.bbc.com/news/world-africa-63988368

Higginson, M., Nadeau, M.-C., & Rajgopal, K. (2019, January). *Blockchain's Occam Problem*, https://www.mckinsey.com/industries/financial-services/our-insights/blockchains-occam-problem?cid=other-eml-alt-mip-mck&hlkid=f1ff7216a70e4041951d60293978a0ea&hctky=2762145&hdpid=95e9bdfa-0709-4b4d-8252-f401bcaac86d

Hollingsworth, J. R., Schmitter, P. C., & Streeck, W. (1994). *Governing capitalist economies*. Oxford University Press.

IBM. (n.d.). *IBM Pricing*, https://www.ibm.com/support/knowledgecenter/bluemix_stage/services/blockchain/howto/pricing.html

Insider Studios. (2022). Here's why Celo project is gaining round and it has a lot to do with DeFi projects and web 3.0, March 24, 2022, https://www.businessinsider.in/cryptocurrency/news/heres-why-celo-project-is-gaining-ground-and-it-has-a-lot-to-do-with-defi-projects-and-web-3-0/articleshow/90420317.cms

Kapadia, S. (2018). Blockchain solution promises to save millions for ocean freight. *SupplyChainDive*, https://www.supplychaindive.com/news/blockchain-ocean-freight-savings/519237/

Kim, H. M., & Laskowski, M. (2018). Toward an ontology-driven blockchain design for supply-chain provenance. *Intelligent Systems in Accounting, Finance and Management, 25*(1), 18–27.

Kong, S. (2020). Blockchain's been a Bust for China's 'Blockchain 50' Public Companies. *Decrypt*, https://decrypt.co/41657/blockchains-been-a-bust-for-chinas-blockchain-50-public-companies

Kshetri, N. (2020). Blockchain-based financial technologies and cryptocurrencies for low-income people: Technical potential versus practical reality. *IEEE Computer, 53*(1), 18–29.

Kshetri, N. (2021). *Blockchain and supply chain management*. Elsevier.

Kshetri, N. (2023). *Fourth revolution and the bottom four billion: Making technologies work for the poor*. University of Michigan Press.

Lacity, M., & Van Hoek, R. (2021). What we've learned so far about blockchain for business. *MIT Sloan Management Review*, https://sloanreview.mit.edu/article/what-weve-learned-so-far-about-blockchain-for-business/

Ledger Insights. (2020). UN World Food Programme uses blockchain for direct payments. *Ledger Insights*, https://www.ledgerinsights.com/un-world-food-programme-uses-blockchain-for-direct-payments/

Lemieux, V. (2017). Evaluating the use of blockchain in land transactions: An archival science perspective. *European Property Law Journal*, https://www.degruyter.com/view/j/eplj.2017.6.issue-3/eplj-2017-0019/eplj-2017-0019.xml

Loadstar. (2020). South Asia Gateway Terminals becomes first Sri Lankan terminal to join TradeLens to digitalise supply chains. *The Load Star*, https://theloadstar.com/south-asia-gateway-terminals-becomes-first-sri-lankan-terminal-to-join-tradelens-to-digitalise-supply-chains/

Macdonald, K. (2007). Globalising justice within coffee supply chains? Fair trade, Starbucks and the transformation of supply chain governance. *Third World Quarterly, 28*, 4, 793–812.

Maersk. (2022). *A.P. Moller—Maersk and IBM to discontinue TradeLens, a blockchain-enabled Global Trade Platform*. https://www.maersk.com/news/articles/2022/11/29/maersk-and-ibm-to-discontinue-tradelens

Mallalieu, K., & Lessey, M. (2012, March 21). Mobile apps boost Trinidad and Tobago fish market. *Digital Opportunity*.

McFerson, H. M. (2009). Governance and hyper-corruption in resource-rich African countries. *Third World Quarterly, 30*(8), 1529–1547.

McMillan, T. (2018). How China plans to feed 1.4 billion growing appetites. *National Geographic*, https://www.nationalgeographic.com/magazine/2018/02/feeding-china-growing-appetite-food-industry-agriculture/

Melcher, F., Sitnikova, M., Graupner, T., Martin, N., Oberthür, T., Henjes-Kunst, F., Gäbler, E., Gerdes, A., Brätz, H. W., Davis, D., & Dewaele, S. (2008). *Fingerprinting of conflict minerals: columbite-tantalite ("coltan") ores.* The Society for Geology Applied to Mineral Deposits No. 23, http://www.e-sga.org/fileadmin/sga/newsletter/news23/SGANews23.pdf

Morris, N. (2019). Hapag-Lloyd, ONE join IBM Maersk TradeLens shipping blockchain. *Ledger Insights,* https://www.ledgerinsights.com/hapag-lloyd-one-ibm-maersk-tradelens-shipping-blockchain/

Namcios. (2021). *LocalBitcoins cuts deposit, transaction fees between wallets in platform to zero.* [online] Bitcoin Magazine: Bitcoin News, Articles, Charts, and Guides. https://bitcoinmagazine.com/business/localbitcoins-red uce-fees-to-zero

NEC Corporation. (2019). NEC, IDB Lab and NGO Bitcoin Argentina to deploy a Blockchain-ba. *Bloomberg,* https://www.bloomberg.com/press-rel eases/2019-08-26/nec-idb-lab-and-ngo-bitcoin-argentina-to-deploy-a-blockc hain-ba

Nguyen, A. (2019). Thai bank targets $1 billion spinoff among its fintech units. *Bloomberg,* www.bloomberg.com/news/articles/2019-11-21/thai-bank-targets-1-billion-spinoff-among-its-fintech-units

Osterfeld, D. (1992). *Prosperity versus planning. How government stifles economic growth.* Oxford University Press.

Pergelova, A., Manolova, T., Simeonova-Ganeva, R., & Yordanova, D. (2019). Democratizing entrepreneurship? Digital technologies and the internationalization of female-led SMEs. *Journal of Small Business Management, 57*(1), 14–39. https://doi.org/10.1111/jsbm.12494

Porter, M. E., & Kramer, M. R. (2002). The competitive advantage of corporate philanthropy. *Harvard Business Review, 80*(12), 56–68.

PYMNTS. (2018, June 25). AliPay, GCash launch blockchain cross-border remittance service. *PYMNTS,* https://www.pymnts.com/news/cross-border-com merce/2018/alipay-gcash-blockchain-cross-border-remittance-philippines/

Raymond, J. (2008). Benchmarking in public procurement. *Benchmarking: An International Journal, 15*(6), 782–793. https://doi.org/10.1108/146357 70810915940

Reuters. (2021). Bitcoin donations surge to jailed Kremlin critic Navalny's cause. *The Jerusalem Post,* https://www.jpost.com/breaking-news/bitcoin-donations-surge-to-jailed-kremlin-critic-navalnys-cause-658766

Romzek, B., LeRoux, K., & Blackmar, J. (2012). A preliminary theory of informal accountability among network organizational actors. *Public Administration Review,* https://onlinelibrary.wiley.com/doi/full/10.1111/j.1540-6210.2011.02547.x?casa_token=wslGT-KbpUAAAAAA%3AJzd3zJOXboPf8 QpAFyFnArhREL16ZgJfoY06Q44C4o2k01aMKFBSOm3B-5YVtNcWo4Lt4 sqVgS7V89B9

Sedex. (2013). Briefing transparency. *Sedex*, https://www.sedex.com/briefing-transparency/

Siedek, H. (2015). Thoughts on DFS in "Europe Minus Infrastructure"—DRC! *MicroSave Consulting*, https://www.microsave.net/2015/07/29/thoughts-on-dfs-in-europe-minus-infrastructure-drc/

Stanley, A. (2017). Microlending startups look to blockchain for loans. *Coindesk*. www.coindesk.com/microlending-trends-startups-look-blockchain-loans.

Suprunov, P. (2018). How much does it cost to hire a blockchain developer? *Medium*, https://medium.com/practical-blockchain/how-much-does-it-cost-to-hire-a-blockchain-developer-16b4ffb372e5

Thorsten, B., Demirgüç-Kunt, A., & Honohan, P. (2009). Access to financial services: Measurement, impact, and policies. *World Bank Research Observer*, *24*(1), 119–145. https://academic.oup.com/wbro/article/24/1/119/167 0507/Access-to-Financial-Services-Measurement-Impact

TradeLens. (2020). Royal Malaysian Customs Department adopts IBM and Maersk's TradeLens blockchain. *TradeLens*, https://www.tradelens.com/press-releases/royal-malaysian-customs-department-adopts-ibm-and-maersks-tradelens-blockchain

UNCTAD. (2020). *UNCTAD policy brief*. https://unctad.org/system/files/official-document/presspb2020d1_en.pdf

United Nations. (2022). LDC identification criteria and indicators. *United Nations*, https://www.un.org/development/desa/dpad/least-developed-country-category/ldc-criteria.html

Webber, D. (2021, April 21). Cryptocurrency in cross-border payments: After coinbase's success, can crypto flourish beyond assets? https://www.forbes.com/sites/danielwebber/2021/04/21/cryptocurrency-in-cross-border-payments-after-coinbases-success-can-crypto-flourish-beyond-assets/?sh=7a02b5 3d416f

West, D. M. (2012). *How mobile technology is driving global entrepreneurship*. Brookings Institution. http://www.insidepolitics.org/brookingsreports/m_e ntrepreneurship.pdf

Wood, M. (2020). Thai customs joins TradeLens blockchain platform. *Ledger Insights*, https://www.ledgerinsights.com/thai-customs-joins-tradel ens-blockchain-platform/

worldbank.org. (2022). Remittances to reach $630 billion in 2022 with record flows into Ukraine, *World Bank*. https://www.worldbank.org/en/news/press-release/2022/05/11/remittances-to-reach-630-billion-in-2022-with-record-flows-into-ukraine

INDEX